89

$ 70.00

AF566892

KEYGUIDE TO INFORMATION SOURCES IN

Aquaculture

KEYGUIDE TO INFORMATION SOURCES IN

Aquaculture

Deborah A. Turnbull

MANSELL

LONDON AND NEW YORK

First published 1989 by
Mansell Publishing Limited, *A Cassell Imprint*
Artillery House, Artillery Row, London SW1P 1RT, England
125 East 23rd Street, Suite 300, New York 10010, U.S.A.

© Deborah A. Turnbull, 1989

All rights reserved. No part of this publication may be reproduced or transmitted in any form or by any means, electronic or mechanical, including photocopy, recording or any information storage or retrieval system, without permission in writing from the publishers or their appointed agents.

The designations employed and the presentation of the material in this publication do not imply the expression of any opinion whatsoever on the part of the author or the publisher concerning the legal status of any country or territory, or of its authorities, or concerning the delimitations of the frontiers of any country or territory.

British Library Cataloguing in Publication Data

Turnbull, Deborah A.
Keyguide to information sources in aquaculture
1. Aquaculture—Bibliographies
I. Title
016.63'09162

ISBN 0-7201-1853-0

Library of Congress Cataloging-in-Publication Data

Turnbull, Deborah A.
Keyguide to information sources in aquaculture / Deborah A. Turnbull.
p. cm.
Bibliography: p.
Includes index.
ISBN 0-7201-1853-0 : $54.00 (US)
1. Aquaculture—Bibliography. 2. Aquaculture—Information services.
3. Aquaculture—Information services—Directories.
I. Title.
Z5970.T87 1989
[SH135]
639-dc19 89—30837
CIP

This book has been printed and bound in Great Britain:
typeset in Baskerville by Colset (Private) Ltd., Singapore,
printed by Redwood Burn Ltd., Trowbridge, on Redwood
Book Wove and bound by WBC Bookbinders.

To aquaculturalists worldwide, my colleagues,
friends and family

Contents

Introduction

Information on aquaculture is proliferating at an enormous rate. Textbooks, journals, newsletters, scientific papers, and directories appear regularly, with new titles coming on to the market every month. Given the amount of information available, the problem lies in finding out what is available and how to access it. Owing to the multidisciplinary nature of the field, those seeking information on aquaculture may encounter formidable obstacles. The problem is not the shortage of information, but how to locate and retrieve it and how to identify which information is appropriate to the users' needs. This *Keyguide* is an attempt to promote awareness of the scope of the resources available, and to assist the user in finding them. It does not duplicate the many directories and bibliographies already in existence; rather, these resources are catalogued in this *Keyguide*. But the *Keyguide* is more than a catalogue: it is a collection of what are considered to be key references, databases, and organizations in aquaculture. The citations given for these sources are designed to provide the user with as much information as possible on the type of resource, the information it offers, and, by means of the source and organizations directories, how to obtain it.

Arrangement of contents

This *Keyguide* is a first source of information. It was designed and prepared to provide an integrated and comprehensive guide to the documentation, reference aids, and key organizational sources of information in aquaculture worldwide.

Part I provides an overview to the field of aquaculture and its literature. Part II is a selected annotated bibliography of major reference sources arranged by document type, for example encyclopedias, dictionaries, directories, key textbooks, bibliographies, atlases, journals, and handbooks. Given the need for up-to-date information, descriptions of aquaculture journals (Section 6.6), databases (Section 8.2) and newsletters (Section 9.1) are provided. Sections 7 and 8 focus on searching the literature (libraries and databases), and Section 9 presents current information sources. Many of the publications listed have been reviewed by the author. The choice of entries has been selective, based on the author's experience. The author hopes that the major sources have been included and apologizes to those authors and editors whose works were obviously omitted as an oversight. As much information as possible has been included about each publication including the language in which it appears and where to obtain a copy of it. A list, with descriptions, of selected aquaculture libraries and organizations worldwide that offer information services is found in Part III.

In a publication of this kind, errors are almost inevitable, and the author will be grateful to readers for pointing out any that have occurred. To report any errors or omissions discovered in the use of this *Keyguide*, please photocopy the entire page on which the error occurs, note the corrections and send to:

AGRODEV CANADA INC.
Suite 600
222 Somerset St. W.
Ottawa, Ontario
Canada K2P 2G3

Acknowledgements

This *Keyguide* has been months in preparation. The person to whom I owe special thanks for her assistance in this preparation is Leigh-Ann Topfer, Corporate Librarian, Agrodev Canada Inc. She has devoted many hours to the compilation of this work.

Thanks also are extended to the World Aquaculture Society and the Aquaculture Association of Canada, who included in their publications notices requesting information, and the staff of Agrodev Canada, who provided assistance and financial support in the preparation of this *Keyguide*.

Special thanks are also extended to Claudine Lemaire of Agrodev Canada Inc., who word-processed the manuscript.

Abbreviations Used for Languages

Al	Albanian
Ar	Arabic
Bu	Bulgarian
Ch	Chinese
Cz	Czech
Da	Danish
De	German
En	English
Es	Spanish
Fi	Finnish
Fr	French
Gr	Greek
Hu	Hungarian
Ic	Icelandic
In	Indonesian
Ir	Irish
It	Italian
Ja	Japanese
Ko	Korean
La	Latin
Nd	Dutch
No	Norwegian
Np	Nepali
Pl	Polish
Pt	Portuguese
Rm	Romanian
Ru	Russian
Sl	Slovak
Sw	Swedish
Th	Thai
Vi	Vietnamese
We	Welsh

Acronyms

ACMRR	Advisory Committee on Marine Resources Research (of FAO)
ADAB	Australian Development Assistance Bureau
ADB	Asian Development Bank
ADC	Agriculture Development Council
ADCP	Aquaculture Development and Coordination Programme (FAO/UNDP)
AFS	American Fisheries Society
AGRIASIA	A quarterly bibliographic journal issued by AIBA
AGRICOLA	Agricultural Online Access (U.S. National Agricultural Library)
AGRINDEX	A monthly bibliographic journal/magnetic tape issued by FAO (AGRIS)
AGRIS	International Information System for the Agricultural Sciences and Technology (of FAO)
AIBA	Agricultural Information Bank for Asia (of SEARCA)
AID	Agency for International Development (U.S.A.)
AIT	Asian Institute of Technology (in Thailand)
APU	AGRIS Processing Unit (in Vienna)
AQD	Aquaculture Department (of SEAFDEC, located in the Philippines)
AQUIS	Aquaculture Information System (of ADCP)
ARDA	*AID Research and Development Abstracts* (issued by USAID)

ASEAN	Association of Southeast Asian Nations
ASFA	Aquatic Sciences and Fisheries Abstracts (issued by FAO/ASFIS)
ASFIS	Aquatic Sciences and Fisheries Information System (of FAO)
ASLS	Aquaculture Scientific Literature Service (of SEAFDEC)
BIOSIS	Bioscience Information Service (U.S.A.)
BIOTROP	SEAMEO Regional Centre for Tropical Biology (Bogor, Indonesia)
BOBP	Bay of Bengal Programme (FAO project)
BRAIS	Brackishwater Information Service
BRS	Bibliographic Retrieval Services (U.S.A.)
CA	*Chemical Abstracts*
CARIS	Current Agricultural Research Information System (of FAO)
CGIAR	Consultative Group on International Agriculture Research
CIDA	Canadian International Development Agency
COFAF	Committee on Food, Agriculture and Forestry (including fisheries), ASEAN
COFI	Committee on Fisheries (of FAO)
COMAR	Integrated Management of Coastal Systems (UNESCO)
CSIRO	Commonwealth Scientific and Industrial Research Organization (Australia)
DANIDA	Danish International Development Agency
EEC	European Economic Community
EIFAC	European Inland Fisheries Advisory Commission (of FAO)
ESCAP	Economic and Social Commission for Asia and the Pacific (of the UN)
FACT	*Freshwater and Aquaculture Contents Tables*
FAO	Food and Agriculture Organization of the United Nations
FSTA	*Food Science and Technology Abstracts* (issued by IFIS)
GCFI	Gulf and Caribbean Fisheries Institute
GEBCO	General Bathymetric Charts of the Oceans (IOC/IHO project)
GEMS	Global Environmental Monitoring System (of UNEP)
GEMSI	Group of Experts on Methods, Standards and Intercalibration (of GIPME)
GESAMP	Joint Group of Experts on the Scientific Aspects of Marine Pollution (IMCO/FAO/UNESCO/WMO/WHO/IEA/UN/UNEP)
GFCM	General Fisheries Council for the Mediterranean

GIPME	Global Investigation of Pollution of the Marine Environment (of IOC)
GTZ	German Agency for Technical Cooperation (Deutsche Gesellschaft für Technische Zusammenarbeit)
IAALD	International Association of Agricultural Librarians and Documentalists
IAMSLIC	International Association of Marine Science Libraries and Information Centers
IARC	International Agricultural Research Centers
IBRD	International Bank for Reconstruction and Development
ICES	International Council for the Exploration of the Sea
ICLARM	International Center for Living Aquatic Resources Management (at Manila, The Philippines)
ICOD	International Centre for Ocean Development
ICMRD	International Center for Marine Resource Development (of URI)
IDRC	International Development Research Centre (Canada)
IFIS	International Food Information Service
IGOSS	Integrated Global Ocean Services System (of IOC)
IMS	*International Marine Science Newsletter* (UNESCO)
INFOFISH	Marketing Information and Advisory Services for Fish Products in the Asian/Pacific Region (FAO project)
INFOTERRA	International Referral System (of UNEP)
IOC	Intergovernmental Oceanographic Commission (of UNESCO)
IPFC	Indo-Pacific Fishery Commission (of FAO)
IRPTC	International Register of Potentially Toxic Chemicals (of UNEP)
JICA	Japan International Cooperation Agency
MCST	*Marine Science Contents Tables*
MEDI	Marine Environmental Data Information Referral System (IOC)
MEDPOL	Mediterranean Pollution Monitoring and Research Programme (of UNEP)
NACA	Network of Aquaculture Centres in Asia (of ADCP)
NAIS	(U.S.) National Aquaculture Information System
NOAA	National Oceanic and Atmospheric Administration (U.S.A.)
NODC	National Oceanographic Data Center (of NOAA)
NORAD	Norwegian Agency for International Development
OECD	Organisation for Economic Co-operation and Development

OETO	Ocean Economics and Technology Office (of the UN)
OSLR	Ocean Science in Relation to Living Resources (an IOC programme)
RAPA	Regional Office for Asia and the Pacific (of FAO)
ROSTSEA	Regional Office for Science and Technology for Southeast Asia (UNESCO, Indonesia)
RS/PAC	Regional Seas Programme Activity Centre (of UNEP)
SAFIS	Southeast Asian Fisheries Information Service (of the SEAFDEC Secretariat)
SCOR	Scientific Committee on Oceanic Research (of ICSU)
SEAFDEC	Southeast Asian Fisheries Development Center
SEAFIS	Southeast Asian Fisheries Information System (a SEAFDEC concept)
SEAMEO	Southeast Asian Ministers of Education Organization
SEAMES	Southeast Asian Ministers of Education Secretariat (of SEAMEO)
SEARCA	Southeast Asian Regional Center for Graduate Study and Research in Agriculture (of SEAMEO)
SIDA	Swedish International Development Authority
SPC	South Pacific Commission (at Noumea, New Caledonia)
TCDC	Technical Cooperation among Developing Countries (a UN concept)
TEMA	Training, Education, and Mutual Assistance (in marine sciences) (IOC)
UNCLOS	United Nations Conference on the Law of the Sea
UNDP	United Nations Development Programme
UNEP	United Nations Environment Programme
UNES	United Nations Environment Secretariat
UNESCO	United Nations Educational, Scientific and Cultural Organization
UNITAR	United Nations Institute for Training and Research
URI	University of Rhode Island (U.S.A.)
USAID	See AID
WHO	World Health Organization
WMO	World Meteorological Organization

PART I

Survey of Aquaculture and Its Literature

Survey of Aquaculture and Its Literature

1 HISTORICAL INTRODUCTION

The word 'aquaculture' derives from the Latin 'aqua' (water) and 'culture' (cultivation, especially with a view to improvement). The culture of organisms in water dates back thousands of years. The Gauls and Romans were known to culture oysters, and artificial fish propagation is known back to the fifth century B.C. when the legendary Chinese Croesus, Fan-li, is said to have reared carp in ponds (Bardach *et al.*, 1972 [99]). (Numbers in square brackets are cross-references to numbered items in Parts II and III.) It is probable that aquaculture dates to much earlier times than this—to civilizations in the Near East that were oriented to water activities and in which fish formed an important part of the diet.

Despite aquaculture's ancient origins, the wild organisms of the oceans, lakes and rivers, and other water bodies have always made a much more significant contribution to the socioeconomic development and diet of cultures and civilizations than has aquaculture. However, given today's increased levels of production from wild fisheries, and the resulting decrease in the size of the stocks, ecological principles dictate that eventually a maximum will be reached in the capture and harvest of wild aquatic organisms. In recognition of this, and in conjunction with increased understanding of the technological, economic, and marketing aspects of cultured aquatic organisms, interest worldwide in the potential social, economic, and nutritional benefits of aquaculture has increased substantially.

The organisms most commonly cultured include fish, shellfish, and algae. The methods of culture vary from the growing of fish in ponds to the culturing of oysters in closed systems using waste heat.

Major constraints on the extensive application of aquaculture worldwide include: (1) difficulties in obtaining a supply of fry of the important species (70 per cent of cultivated food fishes do not breed voluntarily in captivity); (2) lack of information regarding the natural food and artificial feeds required for effective growth and reproduction; (3) difficulties in controlling diseases and parasites; and (4) lack of precise knowledge of the genetics, selective breeding, and hybridization of fish and shellfish. Even with these limitations, however, aquaculture is clearly recognized as an economically viable method of improving the food supplies of the less developed countries and providing value-added products all the year round to developed countries, and the industry is developing so rapidly that many of these constraints are now being overcome.

2 RELATIONSHIP WITH OTHER SUBJECTS

Aquaculture is a multidisciplinary field, as illustrated in Figure 1. Its study involves input from each of the fields shown in Figure 1 and probably, in specific cases, many more. Overlap with other disciplines demands that literature from each of these fields be reviewed in the study of aquaculture or in the practical implementation and operation of aquaculture enterprises.

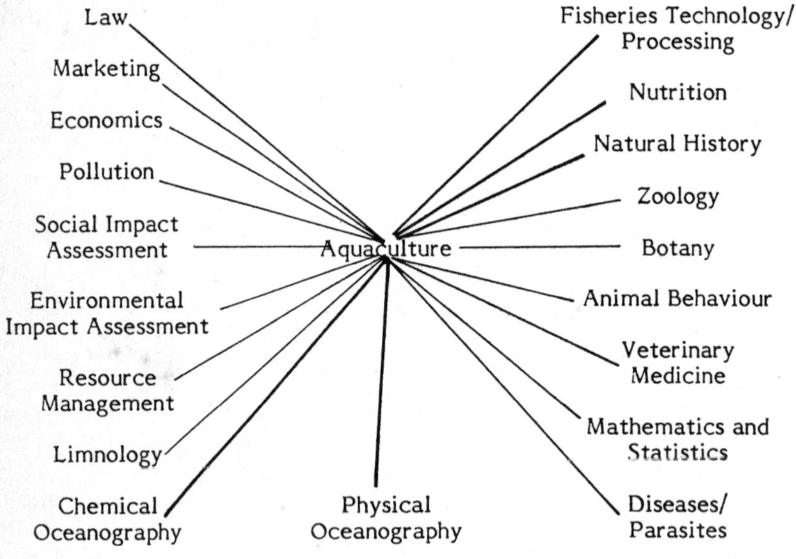

Figure 1 Multidisciplinary nature of aquaculture fisheries

3 ORGANIZATION OF AQUACULTURE

Aquaculture operations are being undertaken by individuals, state farms, commercial undertakings, cooperatives, schools, private companies, and national and regional organizations. Operations vary in scale from some of backyard size to others covering thousands of hectares of land.

Government support for the development of national or localized aquaculture operations has been targeted at research, extension, information exchange, and support for the small-scale operators and subsistence-level farmers. The legal framework is only now becoming formalized in some countries.

The private sector has begun, through aquaculture associations, to have a stronger collective voice in the establishment of the regulations, laws, and policies for local industries.

The number of agencies providing short-term and long-term training programmes in aquaculture is increasing every few months as the demand increases. This has been particularly true in the expansion of the salmon and penaeid shrimp culture industries. The United Nations Food and Agriculture Organization, FAO, has taken the lead in promoting aquaculture development in developing countries, and the World Aquaculture Society and the European Aquaculture Society have played a similar role in developed countries.

The future of aquaculture is extremely bright. Exchange of information concerning new and improved technologies and access to broader markets will be the keys to its success.

4 INTERNATIONAL COOPERATION AND COLLABORATION IN AQUACULTURE INFORMATION EXCHANGE

International cooperation in fisheries research dates back to 1902 when ICES, the International Council for the Exploration of the Sea, was formed. Most of the other organizations involved in fisheries research, development, and promotion were formed during the 1960s or later. They include: EIFAC, the European Inland Fisheries Advisory Commission; IPFC, the Indo-Pacific Fishery Commission; COFI, the Committee on Fisheries of FAO; and GFCM, the General Fisheries Council for the Mediterranean.

In comparison, international cooperation in aquaculture research is in its infancy. However, the situation is improving under the leadership of FAO's Advisory Panel on Aquaculture, the Aquaculture Development and Coordination Programme, and the Aquatic Sciences and Fisheries Information System (ASFIS), and by the promotion of information exchange, cooperation, and collaboration among individual researchers, and exchange of information at international workshops and subject-specific symposia.

Aquaculture development in developing countries has been and is being promoted by various technical assistance organizations including the Swedish SIDA, the Norwegian NORAD, USAID, the Development Banks, the International Science Foundation in Stockholm, Canada's CIDA and IDRC, as well as the German Technical Cooperation Agency.

To date, cooperation on an international scale can be best exemplified by the development of international information systems. The word 'system', as it is used in this context, has been defined as applying not only to the processing of bibliographic information, but also to the arrangements for collecting this information from sources around the world and for its dissemination to users around the world.

At the present time, the existing information systems concerned with fisheries and aquaculture include the Aquatic Sciences and Fisheries Information System (ASFIS), the U.S. National Aquaculture Information System (NAIS), the International Information System for the Agricultural Sciences and Technology (AGRIS), and the Agricultural Information Bank of Asia (AIBA).

ASFIS is the international information system for the science and technology of marine and freshwater environments. It was established in 1971 jointly by FAO and the intergovernmental Oceanographic Commission of UNESCO. The operation of the system is monitored by the Fisheries Department of FAO. ASFIS has twelve participating national centres, eleven of which are located in developed countries. The Centro de Información Científica y Humanística of the Universidad Nacional Autónoma de México is the only centre found in a developing country. Of the others, one is located in Russia, two in Germany, two in France, one in Canada, two in England, one in the United States, one in Portugal, and one at the FAO offices in Rome. ASFIS products include:

1. *Aquatic Sciences and Fisheries Abstracts* (*ASFA*)
2. *Marine Science Contents Tables* (*MSCT*)
3. *Freshwater and Aquaculture Contents Tables* (*FACT*)
4. *World List of Aquatic Sciences and Fisheries Serial Titles*
5. *The International Directory of Marine Scientists*

ASFA is published monthly in two volumes. Part 1 includes biological sciences and living resources abstracts, while Part 2 lists ocean technology, policy and nonliving resources abstracts. Each issue contains 1,000–1,500 abstracts followed by author, subject, taxonomic, and geographic indexes. It has the greatest coverage of aquaculture information of any abstracting journal and ASFIS claims that it includes more than 75 per cent of the world's fisheries literature. Article input is derived from 1,900 serials whose contents are predominantly relevant to the subject scope of *ASFA*. In addition, 3,000 other publications and serials that occasionally publish relevant scientific work are monitored.

At the present time, as pointed out in the Report of the Meeting of an Aquaculture Information Group held in Rome in April 1979, 'a significant proportion of relevant aquaculture information is unavailable [to ASFIS] as much of it is not suited for publication in scientific journals and consequently, remains unidentified in unpublished reports, theses, and similar formats to which access can be gained only by the direct participation of their producers in information exchange programmes at either the national or regional level.' This report goes on to state, 'There are many relatively less known periodicals and occasional publications, especially published in developing countries, that contain considerable information of value.'

This group also noted that the coverage of non-English-language publications in *ASFA* was very incomplete and did not include extremely relevant information from developing countries (which is often written in local languages).

The International Information System for the Agricultural Sciences and Technology (AGRIS) is a worldwide information system that collects and distributes information about documentation to its member countries, and facilitates exchange of information among them. It is coordinated by FAO. Every year some 250,000 documents are generated on ways of improving agriculture and producing more food. Output is available on magnetic tape or in printed form by publication of *AGRINDEX*, a monthly current awareness service which contains, on average, 10,000 citations per issue, of which about one-sixth are nonconventional literature. Input for *AGRINDEX* comes from ninety-two national centres located in developed and developing countries. These centres are predominantly interested in agriculture information but some fisheries information is included in their processing systems. Approximately ten pages of each issue is devoted to aquatic sciences and fisheries information. Relevant subject areas covered include: fisheries production, aquaculture, oceanography, limnology, and aquatic biology.

AGRIASIA is prepared by the Agricultural Information Bank of Asia (AIBA) and published by the Southeast Asian Regional Center for Graduate Study and Research in Agriculture (SEARCA). It provides Southeast Asian regional input to *AGRINDEX* and also includes a significant proportion of extension literature of particular relevance to the region. *AGRIASIA* usually has approximately four pages per issue devoted to aquatic sciences and fisheries citations.

The information scientist involved in collecting and disseminating fisheries information can not only use the resources provided by such international information systems, but also, if the computer software is available, obtain relevant information from some of the commercial databases. A few of these databases include:

a. Biosis Previews (of which *Biological Abstracts* and *Bioresearch Index* are the hard-copy equivalent).
b. SCISEARCH (*Science Citation Index* is the hard-copy equivalent).
c. OAB (hard-copy equivalent *Ocean Abstracts*)

Aquaculture information is also available in journals specific to related subject areas such as medicine, veterinary medicine, agriculture, engineering, oceanography, natural history, and others.

5 PROBLEMS ENCOUNTERED IN OBTAINING AQUACULTURE INFORMATION

Aquaculturists, scientists, researchers, fishermen, librarians, and information scientists encounter the following problems in their search for relevant aquaculture information.

Information Relevance and Availability

When considering the relevance and availability of information, it must be remembered that literature will not solve all the problems of the researcher. This is especially true in a developing country, where even persons who have had appropriate training in investigative development and/or applied research are often frustrated because exact information to answer specific questions, given certain environmental conditions, cannot be obtained. In terms of applying information resulting from research carried out in developed countries to situations in developing countries it has been said that 'It had to be learned (by the developed countries) that scientific-industrial technology, in order to be maximally useful in the underdeveloped countries, could not simply be transferred but had to be adapted to the local prevailing conditions.' This statement has been further simplified in a FAO article which commented that 'It is not that there is insufficient information, but rather a lack of relevant information.'

On the other hand, researchers need to have information available to augment their own training and experience in order to solve problems encountered in fieldwork. The availability of relevant articles is dependent upon the holdings of local universities, institutions, and organizations. In remote areas, away from major university and commercial centres, the problem of information availability worsens. Unless the researcher has access to some type of retrieval service from a library or information scientist, he or she will not be able to obtain information that could be of great assistance. However, even if the holdings of a library are available to the person interested in aquaculture information, a problem in finding relevant material will still exist. The basic difficulty results from the fact that serials holdings of the aquaculture institutions in the developing world are composed predominantly of periodicals published in developed countries, and the articles are about research conducted in the developed countries and, therefore, under temperate conditions.

This point can be illustrated by examining the statistics of the serials holdings of the Southeast Asian Fisheries Development Center (SEAFDEC), which has its aquaculture and fisheries library in Iloilo, the Philippines. Of the 175 journals that SEAFDEC receives by exchange, 82% are published in and had results applicable to developed countries, 10% are published locally, and 8% are received from other developing countries. In terms of serials received by subscription, 94% are published in developed countries, 3% in developing countries, and 3% locally. The analysis of acquisitions lists from other libraries in the developing world shows this to be a familiar pattern.

Local government regulations, the stringent publishing standards of international journals, and the competitive requirement of researchers in the developed world to publish all tend to discourage scientists from developing countries from submitting their research results for publication.

In addition, the aquaculture industry is developing rapidly in many countries and much of the new technology has been developed by the private sector. To provide a forum for exchange of information development by the private sector, countries have organized national and local aquaculture associations to provide a forum for the exchange of information.

Language

The incomplete coverage of non-English publications in international information systems has been noted in a previous section. Expansion of ASFIS may help to solve this problem. However, there is still a need for fisheries institutions to develop effective extension services that should act to make relevant information (particularly technological information) available not only to the researcher involved in either theoretical or applied studies, but also to the first farmer or aquaculturist who may be able to utilize the information and develop methodologies to solve problems. Therefore, there is a need to popularize and translate useful scientific information so that it will be readable and comprehensible to persons not familiar with publications appearing in scientific journals. As a result, the role of the popular science writer and communicator becomes extremely important in the transfer of information to the local fishermen and aquaculturists.

Costs

Exchange of information has been hindered by the costs of building libraries, training personnel, purchasing sophisticated retrieval equipment, postage, travel, publications, books, and journal subscriptions. The costs are further amplified by the value of specific currency in foreign exchange.

The need to try to meet the challenge of fisheries information retrieval and dissemination posed by these problems has been recognized by the International Development Research Centre of Canada (IDRC) and other donors, who have supported and continue to support aquaculture information projects worldwide.

PART II

Bibliography

Bibliography

6 REFERENCE SOURCES

6.1 Encyclopedias

Encyclopedias have played a major role throughout the history of recorded knowledge. Early examples were catalogues of existing knowledge in all areas of science and humanities but were written for a very select audience of scholars. It was not until the nineteenth century that encyclopedias were written for general audiences.

Although most libraries possess encyclopedias, they are an often overlooked source of reference information. They provide concise summaries of the knowledge on a particular subject and can be particularly helpful for a librarian or researcher who has been asked for information on an unfamiliar topic. The encyclopedia article can be an introduction to a topic and also a source of ideas on related subject areas. Other features included in many encyclopedic entries are brief bibliographies for further reading, illustrations, and maps.

Encyclopedias are usually organized in one of two ways: alphabetic or classified. Alphabetic encyclopedias contain brief articles organized alphabetically by subject. A subject/cross-reference index is often included in the final volume. Classified encyclopedias, on the other hand, contain lengthy articles on a subject. Each major article is divided into subtopics that are accessed by way of a detailed subject index. The *Encyclopaedia Britannica*, probably the most famous of all encyclopedias, uses a combination of alphabetic and classified entries. Its first section, the *Micropaedia*, contains concise entries that give an introduction to the subject, and refer the reader to more detailed articles in the second section, the

Macropaedia. The final section of the *Encyclopaedia Britannica* is the *Propaedia*, which is a single-volume overview of existing human knowledge.

General encyclopedias, such as the *Encyclopedia Americana* or the *Encyclopaedia Britannica*, contain useful information on many areas of aquaculture. Further information can be found in the specialized marine science encyclopedias such as those listed in the following bibliography.

1 **Fairbridge, R. W.**, ed. 1966. *Encyclopedia of Oceanography*. New York, NY: Reinhold. 1,021 p.

This one-volume encyclopedia includes concise descriptions, illustrations, photographs, maps, and brief bibliographies on topics related to all areas of oceanography.

2 **George, J. D.** and **J. J. George.** 1979. *Marine Life: An Illustrated Encyclopedia of Invertebrates in the Sea*. New York, NY: Wiley-Interscience. 288 p.

3 *Grande Encyclopédie Alpha de la Mer*. 1972–75. Paris: Grange Batelière. 10 vols.

4 **Groves, D. G.** and **L. M. Hunt.** 1980. *Ocean World Encyclopedia*. New York, NY: McGraw-Hill. 443 p.

This encyclopedia is a nontechnical reference written for students and nonspecialists. Over 400 articles discuss all major areas of oceanographic and marine sciences.

5 *Il Mare: Grande Enciclopedia Illustrata*. 1979–80. Novara, Italy: Istituto Geografico de Agnostini. 10 vols.

6 **Parker, S. P.**, **J. Weil**, and **B. Richman**, eds. 1980. *McGraw-Hill Encyclopedia of Ocean and Atmospheric Sciences*. New York, NY: McGraw-Hill. 580 p.

The marine environment is described in over 200 articles. Entries are arranged alphabetically with cross-references to related subjects.

6.2 Dictionaries, glossaries, lists of acronyms, thesauri

The basic function of a dictionary is to provide an explanation of the meanings of words. Marine-science dictionaries are essentials tools for every aquaculture information centre, while linguistic dictionaries provide a list of terms with their equivalents in one or more other languages.

Glossaries provide definitions to a more select collection of terms from a particular subject area or scientific discipline. Often glossaries are included in a reference text as a guide to the terminology used in that particular work.

Abbreviations and acronyms references allow the user to determine the standard name or full title of, for example, an institution or journal.

Thesauri provide a list of terms together with alternatives of similar meaning, and can be useful for determining the full sense of a term, or as a working aid to help avoid the repetitive use of a single term. Many thesauri also provide a cross-reference system for use in the organization of a collection of documents for reference and retrieval.

7A **Askim, P.** 1962. *Norsk-Engelsk Maritim-Teknisk Ordbok* (Norwegian-English Maritime/Technical Dictionary). Oslo: Grondahl, 1962. 219 p.
Norwegian-English terminology in a technical dictionary first published in 1936.

7B **Askim, P.** 1959. *Engelsk-Norsk Maritim-Teknisk Ordbok* (English-Norwegian Maritime/Technical Dictionary). Oslo: Grondahl, 1959. 137 p.
English-Norwegian terminology. First published 1953.

8 **Baker, B. B., W. R. Deebel**, and **R. D. Geisenderfer**, eds. 1966. *Glossary of Oceanographic Terms*. Washington, DC: U.S. Naval Oceanographic Office. (U.S. Naval Oceanographic Office. Special Publication, 35). 204 p.

9 **Balay, M. A.** 1968. *Glossary of Mareographic Terms*. Buenos Aires: Instituto Panamericano de Geografía e Historia. (Publicación de la Comisión de Cartografía, Comité de Hidrografía, 299). 146 p.

10 **Ben-Yami, M.** 1975. *Russian-English Glossary of Fishing and Related Marine Terms*. Israel Program for Scientific Translations (for the U.S. Department of Commerce, National Oceanic and Atmospheric Administration). 182 p.

11 **Boltovskoy, E.** 1963. *Planktological Dictionary/Diccionario de la Terminología del Plankton Marino*. Buenos Aires: Secretaría Marina, Servicio de Hidrografía Naval. 107 p.
Over 3,800 technical terms used in scientific literature for the principal organisms of marine plankton are defined and the terminology translated into English, Spanish, German, French, and Russian.

12 *Diccionario Técnico Marítimo*. 1980. Oxford: Pergamon Press. 708 p.
Spanish-language technical dictionary.

13 *Dictionary of Japanese Fish Names and Their Foreign Equivalents* (*Nihon-San gyomei daijiten*). 1981. Tokyo: Ichthyologic Society of Japan. 834 p.

14 **Fischer, K.** 1979. *Fachwörterbuch der Meereskunde-Meerestechnik: Deutsch-English, English-Deutsch*. Rostock-Warnemünde, German Democratic Republic: Institut für Meereskunde der Akademie der Wissenschaften der D.D.R. 329 p.

15 **Gorsky, N. N.** and **V. I. Gorskaya.** 1957. *English-Russian Dictionary of Oceanographical Terms*. Moscow: State Publishing Office for Technical and Theoretical Literature. 292 p.

16 **Gorsky, N. N., V. I. Gorskaya,** and **V. K. Salagina.** 1958. *Deutsch-Russisches Wörterbuch für Ozeanographie.* Moscow: Gostechteoretisdat. 306 p.

17 *Guide to Names and Acronyms of Organizations, Activities, and Projects.* 1983. Rome: FAO. (ASFIS Reference Series, 10). 137 p. (Also see Landi, 1979 [22]).

18 **Hollander, N., H. Mestes,** and **C. Roth.** 1980. *Léxico Marinero.* Oxford: Pergamon Press. 128 p.

19 **Hunt, L. M.** and **D. G. Groves,** eds. 1965. *Glossary of Ocean Science and Undersea Technology Terms.* Arlington, VA: Compass Publications. 172 p.
A comprehensive compilation of over 3,500 engineering and scientific terms in English used in the fields of underwater sound, oceanography, marine sciences, underwater physiology, and ocean engineering.

20 *Ichtyologie/Ichthyology.* 1978. Ottawa: Direction Générale de la Terminologie et de la Documentation. (Bulletin de Terminologie/Terminology Bulletin, 161). 351 p.

21 **de Kerchove, R.** 1961. *International Maritime Dictionary.* New York, NY: Van Nostrand Reinhold. 1,018 p.
An encyclopedic dictionary that provides English definitions and their equivalents in French and German.

22 **Landi, G.** 1979. *Initials and Acronyms of Bodies, Activities and Projects Concerned with Fisheries and Aquatic Sciences.* Rome: FAO. (*FAO Fisheries Circular,* 110. Rev. 3). 111 p. (Also see *Guide to Names . . .,* 1983 [17]).

23 *Law of the Sea Terminology.* 1976. Paris: UN. Documentation and Terminology Service. (Terminology Bulletin, 297. Rev. 1. Add. 1). 314 p.
Select terms are translated into Arabic, English, French, Russian, and Spanish.

24 **Lindberg, G. U., A. S. Heard,** and **T. S. Rass.** 1980. *Multilingual Dictionary of Names of Food-Fishes of World Fauna* (Wörterbuch der Namen der marinen Nutzfische der Weltfauna). Leningrad: Nauka. 561 p.

25 *List of Acronyms and Initials for International Organizations and Programmes.* 1980. Paris: UNESCO, Intergovernmental Oceanographic Commission. (IOC/INF, 425). 148 p.

26 **Manzanos, J. A. de.** 1966. *Marine and Fisheries Dictionary/Diccionario de Pesca y Marina.* Tijuana, Mexico: Editorial Pesca y Marina SA, 'Edimar'. Delegación Mexicana.

English–Spanish dictionary of fisheries, maritime, fish technology and navigation terms.

27 **North, J. P.** 1981. *Annotated Acronyms and Abbreviations of Marine Science Related Activities.* Washington, DC: U.S. Department of Commerce. National Oceanic and Atmospheric Administration. 349 p.

28 *Oceanography: List of Terms Relating to Oceanography and Marine Resources.* 1971. New York, NY: United Nations. (Terminology Bulletin, 265). 445 p.
Ocean terminology is translated into English, French, Spanish, Russian, and Chinese.

29 **OECD.** 1978. *Multilingual Dictionary of Fish and Fish Products.* Farnham, Surrey, England: Fishing News Books. 430 p.
This book was published to promote and facilitate international trade in fish products. The second edition of this OECD dictionary, with corrections, translates terms from English into Danish, Dutch, French, German, Greek, Icelandic, Italian, Japanese, Norwegian, Portuguese, Serbo-Croat, Spanish, Swedish, and Turkish. It includes a detailed index in each of these languages.

30 **Paasch, H.** 1974. *Dictionnaire Anglais–Français, Français–Anglais des Termes et Locutions Maritimes* (Dictionary of maritime terms and phrases). Oxford: Pergamon Press. 320 p.

31 **Pring, A. H.** 1970. *A Swahili Nautical Dictionary.* Dar es Salaam, Tanzania: Chuo Cha Uchunguzi Wa Lughaya Kiswahili (Preliminary Studies in Swahili Lexicon, 1).

32 **Ricker, W. E.** 1973. *Russian–English Dictionary for Students of Fisheries and Aquatic Biology.* Ottawa: Department of Fisheries and Oceans. (Bulletin of the Fisheries Research Board of Canada, 183). 428 p.

33 **Rouville, A. de.** 1957. *Dictionnaire Technique Illustré en Six Langues.* Brussels: Association Internationale Permanente des Congrés de Navigation. 272 p.
A dictionary of terms in French, English, German, Spanish, Italian, and Dutch.

34 *Selected Terms in Fish Culture.* 1981. Rome: FAO. (Terminology Bulletin, 19). 106 p.
English, French, and Spanish translations for fisheries terms are defined.

35 **Sullivan, E.** 1980. *Marine Encyclopaedic Dictionary.* Oxford: Pergamon Press.

36 *Thesaurus of Terms for Aquatic Sciences and Fisheries*. 1976. Rome: FAO. (FAO Fisheries Circular, 344). 242 p. (Revisions, in press—to be published by FAO and Cambridge University Press.)

37 *Trilingual Dictionary of Fisheries Technological Terms—Curing*. 1960. Rome: FAO. 85 p.
An English, French, and Spanish technical dictionary.

38 **Tver, D. F.** 1979. *Ocean and Marine Dictionary*. Centreville, MD: Cornell Maritime Press. 358 p.

39 **Vandenberghe, J. P.** and **L. Y. Chaballe**, comps. 1978. *Elsevier's Nautical Dictionary*. Amsterdam: Elsevier. 950 p.
Over 18,000 terms are translated into six languages: Italian, English, French, Spanish, Dutch, and German. Thc major fields from which terms are drawn include shipbuilding, marine technology, navigation, marine transportation, insurance, trade, Law of the Sea, and port terminology.

40 *Vocabulaire de l'Océanologie*. 1976. Paris: Hachette. 431 p.
This publication provides a list of French oceanographic terms and the corresponding English terms as well as a definition in French. It also covers French terms for the zonation of the sea bottom, marine areas, plankton, salinity, pelagic area, and species classification. A Russian index to the terms is included.

6.3 Directories

Directories assist the researcher to locate information about organizations and other researchers worldwide. A directory can be broadly defined as a listing of people or organizations. It may be arranged alphabetically by name, or grouped by subject. Each entry usually consists of an address, and, often, additional information on the area(s) of research, descriptions of the structure and mandate of the institution, its services and staff.

The main problem facing directory users is finding current information. Many directories, particularly personnel directories, are virtually out of date when they are published. Some publishers issue regular supplements, or publish updated directories annually. Though they become obsolete rather quickly, directories are an essential part of library reference collections.

Aquaculturalists have access to a wide variety of directory references. The following bibliography lists the major aquaculture-related directories according to their geographic area of coverage. Marine science, environmental, and fisheries directories also have been included. Many directories specific to certain limited geographical areas (e.g. states or provinces) have not been included.

International Directories

41 **Andrews, S.** and **M. Natkin.** 1983. *World Environment Handbook: A Directory of Government Natural Resource Management Agencies in 144 Countries.* New York, NY: World Environment Center. 130 p.
An international 'who's who' of the environment field. The environment and natural resource agencies of 144 governments are listed with their addresses and the names of the officials in charge.

42 *Annotated Directory of Intergovernmental Organizations Concerned with Ocean Affairs.* 1976. New York, NY: UN. (A/CONF.62/L.14). 165 p.
Includes fifty-eight global, regional and subregional intergovernmental organizations working in the field of marine affairs. It provides an excellent background document to the United Nations system and its agencies supporting ocean affairs programmes.

43 *Annotated Directory of Intergovernmental Organizations Dealing with Oceanic Matters.* 1977. Rome: FAO Fisheries Department.

44 *Annuaire International du Monde Sous-Marin.* 1983. Paris: Confédération Mondiale des Activités Subaquatiques. 400 p.

45 *Aquaculture Aid Profiles: A Summary of Information on Aquaculture Projects with External Aid Components.* Quarterly. Rome: FAO. Paging varies.
Provides detailed descriptions of aquaculture projects supported by international, regional, and bilateral funding agencies.

46 **Bekiashev, K. A.** and **V. V. Serebriakov**, comps. 1981. *International Marine Organizations: Essays on Structure and Activities.* 1981. Dordrecht, The Netherlands: Martinus Nijhoff. (Developments in Transport Studies, Volume 3.) 578 p.
Includes a listing and description of international maritime organizations, international fisheries organizations, and international organizations concerned with marine sciences.

47 **Caddy, J. F.**, ed. 1982. *Provisional World List of Computer Programmes for Fish Stock Assessment and Their Availability by Country and Fisheries Institute.* Rome: FAO. 51 p.
An annotated catalogue of computer programs in fish stock assessment which are available to outside users.

48 *Coral Reef Directory.* 1986. Cambridge, England: International Union for Conservation of Nature and Natural Resources.

49 *Directory of FAO Fisheries Field Experts.* Quarterly. Rome: FAO. Paging varies.
Current FAO projects and their addresses are given. The arrangement is alphabetical by country.

50 *Directory of Fishing Technology Institutions and Services.* 1980. Rome: FAO. (FAO Fisheries Technical Paper, 205). 99 p.

This trilingual (En, Fr, Es) directory details special activities, facilities, training courses, and subject areas of major institutions worldwide.

51 *Fisheries Department Field Projects*. 1984. Rome: FAO. (Paging varies).

52 **Harvey, N.**, ed. 1983. *Agricultural Research Centres: A World Directory of Organizations and Programmes*. Detroit, MI: Gale Research Company. 2 vols., 1,276 p.

A brief listing of government, private, and academic groups engaged in various forms of renewable resources research. Alphabetical arrangement by country gives addresses and scope of research for some 9,000 organizations and programmes. Institutions involved in fisheries, aquaculture, and food science are included.

53 *International Association of Marine Science Libraries and Information Centers. Membership Directory, 1984–1985*. 1985. Woods Hole, MA: Woods Hole Oceanographic Institution. 8 p.

The membership of IAMSLIC listed by name and institution. The majority of members are from North America, but member libraries in Europe, Australia, South America, and Southeast Asia are also included.

54 *International Directory of Fish Inspection and Quality Control Institutes*. 1984. Rome: FAO. (FAO Fisheries Technical Paper, 244). 139 p.

The objectives and facilities of major institutions are described with the aim of furthering cooperation between these agencies. This directory includes sixty-one institutes from fifty countries worldwide involved in fish inspection and quality control.

55 *International Directory of Fish Technology Institutes*. 1980. Rome: FAO. (FAO Fisheries Technical Paper, 152, Rev. 1). 124 p.

Ninety-four institutes from fifty-five countries are listed. The information included in each entry is the name of the director, address, major fields of interests, facilities and publications.

56 *International Directory of Marine Scientists*. 1983. Paris: UNESCO. 488 p. (text), 173 p. (index).

A directory to over 2,500 institutions and 18,000 specialists in marine science. Three indexes give access by name of institution (subdivided by country), name of specialist, and subject area.

57 *Meteorological Services of the World*. 1982. Geneva: World Meteorological Organization. (WMO/OMM No. 2). 188 p.

The meteorological services of each of the eighty-two member countries of the WMO are described. Each description includes information on general aeronautical and meteorological publications.

58 **Varley, A.**, ed. 1978. *Who's Who in Ocean and Freshwater Science*. Harlow, Essex, England: Longman. 336 p.

59 **Verrel, B.** and **Opitz, H.** 1984. *World Guide to Scientific Associations and Learned Societies*. Munich: Saur. (*Handbook of International Documentation and Information*, Vol. 13). 947 p.

60 **Winn, C. P.**, comp. 1981. *Directory of Marine Science Libraries and Information Centers*. Woods Hole, MA: Woods Hole Oceanographic Institution, Sea Grant Program and International Association of Marine Science Libraries and Information Centers. Unpaged.

A 1984 supplement updates this useful listing, by country, of major governmental and institutional marine science information centres. An index to institutions and librarians simplifies access. Addresses, user policies, collection size and publications are given for each entry.

61 *World Directory of Institutions Concerned with Residues of Agriculture, Fisheries, Industries*. 1982. Rome: FAO. (FAO Agricultural Services Bulletin No. 21, Rev. 2). 219 p.

Contains approximately 1,000 names from 120 countries and aims at facilitating communication and exchange of information between similarly orientated institutions.

62 *World Environmental Directory*. 1983. Silver Springs, MD: Business Publishers International.

An index to the key people and organizations in the environmental field. Over 40,000 entries with names, addresses, and phone numbers.

63 *Yearbook of International Organizations* 1983. Brussels: Union of International Associations.

Regional Directories

Africa

64 **Coche, A. G.** and **P. Eric.** 1979. *List of Aquaculture Personnel in Africa for 1980*. Rome: FAO. Fishery Resources and Environment Division.

65 *Marine Research Centres: Africa, 1982*. 1982. Rome: FAO (for UNEP).

The major marine science institutions and organizations in the following countries are described in this directory: Algeria, Angola, Benin, Congo, Djibouti, Egypt, the Gambia, Ghana, Guinea, the Ivory Coast, Kenya, Libya, Madagascar, Mauritania, Morocco, Mozambique, Nigeria, Senegal, Seychelles, Sierra Leone, Sudan, Tanzania, Tunisia, and Zaire.

Caribbean

66 *Directory of Caribbean Marine Research Centres*. 1980. Geneva: United Nations Environment Programme. Various paging.

A directory of marine research centres in the Caribbean. Non-Caribbean countries which have established centres in the area, for example France and the United States, are also included.

Europe

67 *Directory of Mediterranean Marine Research Centres.* 1979. Nairobi: United Nations Environment Programme. Various paging.

Provides reliable up-to-date information on 144 research centres participating in the UNEP-coordinated Mediterranean Pollution Monitoring and Research Programme (MEDPOL). Information provided for each centre includes: address, name and title of executive officer, scope of research activities, brief history, mission and purpose, organizational structure, three-year review of activities, major current and future activities, cooperative programmes, library, publications and instructional programmes.

North America

68 *Directory of North American Fisheries Scientists.* 2nd ed., 1987. Edited by B. D. McAleer. Bethesda, MD: American Fisheries Society. 363 p.

A directory listing over 8,000 specialists involved in the fields of fisheries and aquaculture. Addresses and telephone numbers are given and a subject speciality and geographic index are included.

69 *Fish Culturist Registry.* 1982. Bozeman, MA: American Fisheries Society. Fish Culture Section.

A listing of the membership of the AFS, Fish Culture Section, with a brief biographical entry for each member and an outline of his or her area of specialization. A keyword index is included.

70 **Watkins, M. M.** and **J. A. Ruffner.** 1983. *Research Centers Directory.* Detroit, MI: Gale Research Company. 1,082 p.

A guide to university-related and other nonprofit research organizations in North America. Fisheries and oceanographic institutions are covered in the section on conservation. The information given includes addresses, contact people, the mandate of the organization, and a description of their field(s) of research and publications.

South and Central America

71 *Directory of Marine Sciences in South America (Directorio de Ciencias del Mar de América del Sur, con excepción de Guyana, Surinam y la Guyana Francesa).* 1980. Montevideo: UNESCO. 258 p. (text), 46 p. (supp.).

A list of the major marine institutions and scientists in South America. Information on each institution includes: areas of work; branches of marine science studied; size of the library; degrees and diplomas awarded; journals published and training courses offered.

Asia

72 *Directory of Indian Ocean Marine Research Centres.* 1978. Geneva: United Nations Environment Programme. Various paging.

73 *Directory of Scientists and Technologists in the Fields of Fisheries Research, Development and Management in the Southeast Asian Region.* 1979. Samutprakarn, Thailand: Southeast Asian Fisheries Development Center.

A directory of specialists, including brief biographies of each, with indexes by name, country, and area of specialization.

74 *NACA: Network of Aquaculture Centres in Asia.* [1980]. Bangkok: United Nations Development Programme. NACA Project.
Describes the work, organization, and sources of funding of NACA, and of each of the national centres in Thailand, India, China, and the Philippines.

75 *Organizations/Agencies Concerned with Fisheries Development in Southeast Asia.* 1984. Bangkok: Southeast Asia Fisheries Development Center. (SEC/SP/9). 126 p.
This directory includes a description of the aims, assistance programmes, and activities of twenty-eight national, regional, and international organizations and development banks that support or contribute to fisheries development in Southeast Asia.

76 *Preliminary Directory of ASEAN Aquaculture Research and Training Institutions.* 1979. Manila: FAO/UNDP. South China Sea Fisheries Development and Coordinating Programme. (ASEAN77/FA.EgA/Rpt.4). 72 p.

National Directories

Canada

77 *Canadian Aquaculture Directory.* 1984. St. Andrews, New Brunswick: Atlantic Salmon Federation.
This directory includes the names and addresses of people involved in aquaculture throughout Canada.

78 *Canadian Fisheries and Ocean Industries Directory, 1984–85.* 1986. Halifax, Nova Scotia: Management Consultants Ltd.

79 *Canadian Fishing and Marine Directory.* Annual. Ottawa: Canadian Fishing Report.
This industry directory includes the names and addresses of key personnel, government agencies and programmes, industrial organizations, and associations.

80 *Directory of Canadian Environmental Experts.* 1984. Ottawa: Canadian Government Publishing Centre. 426 p.
Addresses and a brief description of the fields of research of those included are listed in this update of the 1980 directory.

81 *Directory of Marine and Freshwater Scientists in Canada.* 1986. Ottawa: Department of Fisheries and Oceans. (Canadian Special Publication in Fisheries and Aquatic Sciences, 73R). 259 p.
Lists names and addresses of government, academic, and industrial specialists and organizations in the fields of marine and freshwater science in Canada. Biennial updates.

82 *Fish and Seafood Products Directory*. 1982. Ottawa: Fisheries Council of Canada. 117 p.

China

83 *A Brief Introduction to the Chinese Institutions of Ocean Science and Technology*. 1984. En, Ch. Beijing: China Ocean Press. 240 p.
Describes the services and interests of over forty marine institutions in China.

Colombia

84 **Marzo, D. E.**, comp. 1981. *Directory of Sources of Information on Oceanography*. Bogotá: Fondo Colombiano de Investigaciones Científicas y Proyectos Especiales. Directorio Unidades de Información Ciencias del Mar.

France

85 *Océanologie: Annuaire Technique et Industriel, 1983–84*. 1983. Paris: ASTEO Groupement Interprofessionnel pour l'Exploitation des Océans. 92 p.

Indonesia

86 **Wiraatmadja, N.** and **P. M. North**, comps. 1982. *Indonesia National Directory of Environmental Information Sources*. Jakarta: Indonesian Institute of Sciences.
A guide to Indonesian governmental and nongovernmental organizations concerned with the environment. Addresses, subject and geographic areas, and the types of publications of each agency are given.

United Kingdom

87 **Simpson, P.** and **D. S. Moulder**, comps. 1980. *Directory of Library and Information Facilities*. Plymouth: Marine Biological Association.
Describes the marine biology, oceanography, and fisheries information services available in the United Kingdom.

United States

88 *American Fisheries Directory and Reference Book*. 1981. Camden, ME: National Fisherman. 750-plus p.
A useful source of information on the North American fishing industry and on international marine science. Chapters include a list of international fisheries institutions and organizations, and a bibliography of handbooks and periodicals related to oceanography and fisheries research.

89 *American Fisheries Society Membership Directory and Handbook*. 1984. Bethesda, MD: American Fisheries Society. 98 p.
Membership of the AFS listed by name and by region. Includes information on the AFS code of practices, lists of officers, and annual meetings.

90 *Aquaculture Research: A Directory of USDA and State Projects in CRIS*. 1983. Beltsville, MD: U.S. Department of Agriculture. Library. Current Research Information System, Cooperative State Research Service. 357 p.

An index, by subject, institution, title, and principal researcher, to U.S. government-supported research in aquaculture.

91 **Bryant, J.**, ed. 1985. *Conservation Directory*. Washington, DC: National Wildlife Federation. 302 p.
An annual listing of organizations, agencies, and officials concerned with natural resource use and management.

92 *Directory of Aquaculture Information Resources*. 1982. Beltsville, MD: U.S. Department of Agriculture. National Agricultural Library. (Bibliographies and Literature of Agriculture, 25). 53 p.
A directory of state and national agencies concerned with aquaculture, with a section on primary literature sources, journals, and directories. The index allows access by subject and by species.

93 *List of National Fish Hatcheries and Fishery Assistance Stations*. 1981. Washington, DC: U.S. Department of the Interior. Fish and Wildlife Service. (Fishery Leaflet, 147). 32 p.
Addresses and key personnel are given for each of the national fish hatcheries and for the regional offices of the Fish and Wildlife Service.

94 *List of State Fish Hatcheries and Rearing Stations*. 1976. Washington, DC: U.S. Department of the Interior. Fish and Wildlife Service. 28 p.
A listing of state hatcheries and their addresses, with information on the species propagated at each.

95 *Major Aquaculture Associations, Education and Research Resources in the United States*. 1983. Beltsville, MD: U.S. Department of Agriculture. National Agricultural Library. (Bibliographies and Literature of Agriculture, 26). 174 p.
A listing of state and national aquaculture organizations and associations, including university programmes and publications, and an index of key people in the field.

96 **Renz, L.** 1985. *The Foundation Directory*. New York, NY: The Foundation Center. 885 p.
This annual directory is a useful source of information on nonprofit organizations involved in oceanographic research in the United States.

97 *United States Directory of Marine Scientists*. 1982. Washington, DC: National Academy Press.
A directory of specialists that includes a section listing major organizations, universities, corporations, and government agencies involved in marine science.

6.4 Key textbooks

This section includes general textbooks that provide a basic introduction to aquaculture as a whole, dealing, for example, with its history, development, theory, procedures, and methods of collecting and analysing data and information and the interpretation of results.

98 **Anderson, L. G.**, ed. 1981. *Economic Analysis for Fisheries Management Plans.* Lancaster, PA: Technomic Publishing. 328 p.

98A **Bardach, J. E.** and **J. H. Ryther**. 1968. *The Status and Potential of Aquaculture.* Springfield, VA: National Technical Information Service.
A selective examination of various species of organisms and the methods for cultivation.

99 **Bardach, J. E., J. H. Ryther**, and **W. O. McLarney**. 1972. *Aquaculture: The Farming and Husbandry of Freshwater and Marine Organisms.* New York, NY: Wiley-Interscience. 868 p.
This excellent reference text provides an overview of present, and to some extent past, practices of food aquaculture around the world.

100 **Barnes, R. S. K.** 1980. *Coastal Lagoons.* Cambridge, England: Cambridge University Press. (Cambridge Studies in Modern Biology, No. 1). 130 p.
Provides an introduction to the types, formation, circulation, biology, and use of lagoons.

101 **Barnes, R. S. K.** and **R. N. Hughes.** 1982. *An Introduction to Marine Ecology.* Oxford: Blackwell Scientific. 348 p.

102 **Barnes, R. S. K.** and **K. H. Mann**, eds. 1980. *Fundamentals of Aquatic Ecosystems.* Oxford: Blackwell Scientific. 240 p.

102A **Bazigos, G. P.** 1977. *Mathematics for Fishery Statisticians.* Rome: FAO. (FAO Fisheries Technical Paper, 169). 193 p.
The purpose of this manual is to cover those topics of mathematics required by fishery statisticians and workers in related fields. The manual is divided into three main parts: algebra, calculus, and trigonometry.

103 **Beer, T.** 1983. *Environmental Oceanography: An Introduction to the Behaviour of Coastal Waters.* Oxford: Pergamon Prcss. 109 p.
An explanation of coastal-zone processes, both onshore and offshore, including: coastal engineering, waves and tides, boundary layers, mixing, meteorology, reefs and estuaries, and remote sensing applications.

104 **Bell, F. W.** 1978. *Food from the Sea: The Economics and Politics of Ocean Fisheries.* Boulder, CO: Westview Press. (Special Studies in Natural Resources and Energy Management series).
This reference could serve as a textbook for courses in fishery economics. Biological and economic principles are explained providing a comprehensive analysis of how economic and political factors interact in fisheries management.

105 **Bergman, H. L., R. A. Kimerle**, and **H. W. Maki.** 1985. *Environmental Hazard Assessment of Effluents*. Oxford: Pergamon Press. 390 p.
Presents a complete overview of water quality problems caused by effluents. Covers biological effects, explores assessment, and provides case histories.

106 **Bolis, L., J. Zadunaisky**, and **R. Gilles.** 1984. *Toxins, Drugs and Pollutants in Marine Animals*. Berlin: Springer-Verlag. (Proceedings in Life Sciences series). 200 p.

107 **Boney, A. D.** 1975. *Phytoplankton*. London: Edward Arnold; Baltimore, MD: University Park Press. (The Institute of Biology's Studies in Biology, 52). 124 p.
Both freshwater and marine phytoplankton are represented in this conceptual account of basic knowledge and current research. Illustrated with examples from around the world.

108 **Bowden, K. F.** 1983. *Physical Oceanography of Coastal Waters*. New York, NY: John Wiley. 302 p.
This text is designed for undergraduate and graduate students of oceanography and ocean engineering. The chapters focus on tides, currents, waves, upwelling, salinity, temperature, thermocline, mixing, and circulation patterns.

109 **Boyd, C. E.** 1979. *Water Quality in Warmwater Fish Ponds*. Auburn, AL: Auburn University. 359 p.
This text discusses the principles of water quality, water quality management, and the measurement of water quality in warmwater fish ponds.

110 **Brown, E. E.** 1983. *World Fish Farming: Cultivation and Economics*. Westport, CT: AVI Publishing. 391 p.
Provides a summation of the present state of aquaculture worldwide. The methods used in growing fish and shellfish in thirty-one countries are described in detail, including feeding practices, food yield, and types of production facilities for freshwater, brackish-water, and marine production. Pond raceway, fjord cage, net, and other culturing methods are discussed.

111 **Brown, E. E.** and **J. B. Gratzek.** 1980. *Fish Farming Handbook*. Westport, CT: AVI Publishing. 391 p.
A major text on aquaculture methods and techniques.

112 **Chapman, V. J.** and **D. J. Chapman.** 1980. *Seaweeds and Their Uses*. London: Chapman & Hall. 334 p.
Describes seaweeds, particularly the uses of seaweed.

113 **Chaston, I.** 1983. *Marketing in Fisheries and Aquaculture*. Farnham, Surrey, England: Fishing News Books. 143 p.
This book advocates improvement in the business management capabilities of industry personnel and public sector advisory services, especially in marketing. Topics covered include marketing strategy, buyer behaviour, price

determination, distribution, promotion, personal selling, and control systems.

114 Chaston, I. 1984. *Business Management in Fisheries and Aquaculture*. Farnham, Surrey, England: Fishing News Books. 136 p.
The factors for successful business management, financing, marketing, production, and personnel management are described, with examples from the fishing and aquaculture industries.

115 Clucas, I. J. and **P. J. Sutcliff.** 1981. **Introduction to Fish Handling and Processing**. London: Tropical Products Institute. (Report G. Tropical Products Institute). 143 p.

116 Connell, J. J., ed. 1980. *Advances in Fish Science and Technology*. Farnham, Surrey, England: Fishing News Books. 528 p.
The eighty papers in this text comprise the proceedings of a Torry Research Station conference of world specialists in the field of fisheries technology.

117 Couve, A. 1977. *Aquaculture in Latin America*. Washington, DC: Inter-American Development Bank. (IDB/FAO Cooperative Program INF.8/77).
A study of the potential for and problems involved in developing aquaculture in Latin America. The descriptions for each country include an overview of the current status, a list of research centres and projects, and a summary of the most promising species.

117A Cunningham, S., M. R. Dunn, and **D. Whitmarsh.** 1985. *Fisheries Economics: An Introduction*. London: Mansell; New York: St. Martin's Press. 384 p.
Includes a chapter on aquaculture.

118 Cushing, D. H. 1968. *Fisheries Biology: A Study in Population Dynamics*. Madison, WI: University of Wisconsin Press. 320 p.
A textbook based on a series of lectures on fisheries biology presented at the University of Wisconsin. Topics covered include fisheries biology, fish migration, stock assessment, and Law of the Sea.

119 Cushing, D. H. 1975. *Marine Ecology and Fisheries*. Cambridge, England: Cambridge University Press. 228 p.
An introductory text to marine ecology and fisheries.

120 Cuyvers, L. 1984. *Ocean Uses and Their Regulation*. New York, NY: John Wiley. 179 p.

121 *Developments in Aquaculture and Fisheries Science*. 1976–. Amsterdam: Elsevier Science Publishing.
Each volume in this series discusses a different subject in fisheries and aquaculture:

Vol. 1 *Farming Marine Organisms Low in the Food Chain.*
Vol. 2 *Farming the Cupped Oysters of the Genus* Crassostrea.

Vol. 3 *Farming the Flat Oysters of the Genus* Ostrea.
Vol. 4 *Farming Marine Fisheries and Shrimps.*
Vol. 5 *Sound Reception in Fish.*
Vol. 6 *Disease Diagnosis and Control in North American Aquaculture.*
Vol. 7 *Mussel Culture and Harvest: A North American Perspective.*
Vol. 8 *Chemoreception in Fishes.*
Vol. 9 *Water Quality Management for Pond Fish.*
Vol. 10 *Giant Prawn Farming.*
Vol. 11 *Modern Methods of Aquaculture in Japan.*
Vol. 12 *Genetics in Aquaculture.*
Vol. 13 *Bioeconomics of Aquaculture.*

122 Dring, M. J. 1983. *The Biology of Marine Plants.* London: Edward Arnold; Baltimore, MD: University Park Press. 208 p.
A modern, analytical account of marine plants and their environments around the world. Contains over 250 references covering recent analytical and experimental work.

123 Eddie, G. C. 1984. *Engineering, Economics and Fisheries Management.* Farnham, Surrey, England: Fishing News Books. 108 p.
An examination of the relationships between technology and fisheries management and development.

124 Eisler, R. 1981. *Trace Metal Concentrations in Marine Organisms.* Oxford: Pergamon Press. 685 p.

125 *FAO Species Identification Sheets for Fishery Purposes.* 1973. Rome: FAO.
Mediterranean and Black Sea: Fishing Area 37. 2 vols.
Eastern Indian Ocean: Fishing Area 57.
Western Central Pacific: Fishing Area 71. 4 vols.
Western Central Atlantic: Fishing Area 31. 7 vols.
Eastern Central Atlantic: Fishing Areas 34, 47. 7 vols.
Western Indian Ocean: Fishing Area 51. 6 vols.

126 Fincham, A. A. 1984. *Basic Marine Biology.* Cambridge, England: Cambridge University Press, for the British Museum (Natural History). 157 p.

127 Friedrich, H. 1969. *Marine Biology: An Introduction to Its Problems and Results.* Seattle, WA: University of Washington Press. 474 p.
This text, originally in German, provides an introductory analysis of marine biology including discussions on the morphology of the ocean, ecological factors, animal and plant associations, and distribution.

128 Gerlach, S. A. 1981. *Marine Pollution: Diagnosis and Therapy.* Berlin: Springer-Verlag. 218 p.
Provides a broad view of fundamental problems of marine pollution, evident in all world oceans. Included are discussions of domestic effluents, industrial effluents,

oil pollution, radioactivity, general problems of harmful substances in the sea, global contamination with heavy metals and chlorinated hydrocarbons, laws against pollution of the oceans, and diagnosis and therapy.

129 **Gorlenko, V. M., G. A. Dubinina**, and **S. I. Kuznetsov**. 1983. *The Ecology of Aquatic Micro-Organisms*. Stuttgart: E. Schweizerbart'sche Verlagsbuch. (Die Binnengewasser, Band 28). 252 p.
This is a general text on the morphological diversity of micro-organisms and their role in the cycle of substances in lakes. New methods of investigation are discussed and information is presented on organisms to which little attention has been paid in past research.

130 **Gulland, J. A.**, comp. 1970. *The Fish Resources of the Oceans*. Rome: FAO. (FAO Fisheries Technical Paper, 97). 255 p.
A survey of world fisheries including maps, stock assessments, oceanographic data, resource potential, catch statistics, and bibliographies by geographical region.

131 **Hauser, H.** 1984. *Book of Fishes*. Locust Valley, NY: Pisces Books International Inc. (Skin Diver Magazine's Book of Fishes). 192 p.
Informative guide on the life habits of fish most commonly seen by divers and snorkelers. Includes a discussion of underwater photo techniques.

132 **Heikoff, J. M.** 1980 *Marine and Shoreland Resources Management*. Lancaster, PA: Technomic Publishing. 224 p.
This text serves as an introduction to coastal-zone management including resources, recreation, water quality, and legislative management.

133 **Hildreth, R. G.** and **R. W. Johnson.** 1983. *Ocean and Coastal Law*. Englewood Cliffs, NJ: Prentice-Hall. 560 p.
Concentrates on examples from the experience of the United States; however, it is an excellent reference text to the legal and regulatory issues that all coastal nations must face. The following topics are discussed with detailed examples: (1) private rights, public rights, beach access; (2) fishing management; and (3) living resource management.

134 **Huet, M.** and **J. A. Timmermans.** 1983. *Textbook of Fish Culture: Breeding and Cultivation of Fish*. Farnham, Surrey, England: Fishing News Books. 454 p.
This text provides an introduction to fish breeding in both fresh and brackish water worldwide. Methods of rearing over 100 different species of fish are discussed.

135 **Iversen, E. S.** 1976. *Farming the Edge of the Sea*. Farnham, Surrey, England: Fishing News Books. 440 p.
The text includes a study of aquaculture methods in various regions of the world including a look at the major commercial species farmed. An extensive bibliography is included.

136 Kinne, O., ed. 1977–. *Marine Ecology: A Comprehensive, Integrated Treatise on Life in Oceans and Coastal Waters*. New York, NY: John Wiley.
Vol. 1 *Environmental Factors* (3 pts.).
Vol. 2 *Physiological Mechanisms* (2 pts.).
Vol. 3 *Cultivation*.
Vol. 4 *Dynamics*.
Vol. 5 *Ocean Management* (4 pts.).

137 Kinne, O., ed. 1980–. *Diseases of Marine Animals*. New York, NY: John Wiley.
Vol. 1 *General Aspects: Protozoa to Gastropoda*.
Vol. 2 *Introduction: Bivalvia to Scaphopoda*.

138 Kungvanky, P. 1984. *Overview of Penaeid Shrimp Culture in Asia*. NACA/RLCP Research Document No. 4. UNDP/FAO/NACA. The Philippines.
This report provides an excellent brief overview to penaeid shrimp culture in Asia.

139 Lackey, R. T. and L. A. Nielsen. 1980. *Fisheries Management*. New York, NY: John Wiley. 422 p.

140 Laevastu, T. and **M. L. Hayes.** 1981. *Fisheries Oceanography and Ecology*. Farnham, Surrey, England: Fishing News Books. 216 p.
This general text includes discussions of fisheries ecology, effects of environment factors on fish and the behaviour of stocks, migration patterns of fish, fisheries in relation to environmental factors, the variability of the ocean environment, and numerical ecosystem simulations. This text will benefit fisheries biologists and students as well as fishermen and fisheries managers and administrators.

141 Lauff, G. H. 1967. *Estuaries*. Washington, DC: American Association for the Advancement of Science. (Publication 83). 757 p.
This text comprises a collection of papers on the basic features, physical considerations, geomorphology, sedimentology, microbiology, nutrients, ecology, fisheries, and human influences on estuaries.

142 *The Law of the Sea. Official Text of the United Nations Convention on the Law of the Sea with Annexes and Index*. 1983. New York, NY: United Nations. 224 p.
This volume contains the official text of the UN Convention on the Law of the Sea, signed at Montego Bay, Jamaica, on 10 December 1982. This text is supplemented by a subject index to the 320 articles of the Convention, its six annexes and the four associated resolutions.

143 Laws, E. A. 1981. *Aquatic Pollution: An Introductory Text*. New York: John Wiley. 482 p.
An introduction to aquatic pollution.

144 Lawson, R. 1984. *Economics of Fisheries Development*. London: Frances Pinter. 296 p.

An introductory text on the theoretical principles of fisheries economics and management and how these principles relate to the fisheries of developing countries.

145 Levinton, J. S. 1982. *Marine Ecology*. Englewood Cliffs, NJ: Prentice-Hall. 543 p.

146 Little, D. and **J. Muir.** 1987. *A Guide to Integrated Warmwater Aquaculture*. Stirling, Scotland: Institute of Aquaculture Publications, University of Stirling. 237 p.
This book describes the principles of integrated aquaculture, particularly in ponds in the tropics and illustrates the means by which such farming practices can be developed successfully.

147 Long, S. P. and **C. F. Mason.** 1983. *Saltmarsh Ecology*. Glasgow: Blackie. 168 p.

148 Lowe-McConnell, R. H. 1977. *Ecology of Fishes in Tropical Waters*. London: Edward Arnold; Baltimore, MD: University Park Press. (The Institute of Biology's Studies in Biology, 76). 64 p.
This text presents a full discussion of research on tropical fishes, including community development, behaviour and evolution, commercial exploitation, and conservation.

149 McConnaughey, B. H. and **R. Zottoli.** 1983. *Introduction to Marine Biology*. 4th ed. St. Louis, MO: C. V. Mosby. 638 p.
Its six chapters include such topics as the chemistry of seawater; terrestrial maritime communities; coastal wetlands and sand dunes, benthic communities of coral reefs and exposed rocky areas; deep-sea benthos; and Law of the Sea.

150 Mackenzie, W. C. 1983. *An Introduction to the Economics of Fisheries Management*. Rome: FAO. (FAO Fisheries Technical Paper, 226). 41 p.

151 Mann, K. H. 1982. *Ecology of Coastal Waters: A Systems Approach*. Oxford: Blackwell Scientific. (Studies in Ecology, volume 8). 300 p.

152 Martin, R. E., G. J. Flick, C. E. Holland, and **D. R. Ward**, eds. 1982. *Chemistry and Biochemistry of Marine Food Products*. Westport, CT: AVI Publishing. 473 p.
This text represents the first attempt to draw together, in one volume, present knowledge concerning the chemistry and biochemistry of marine food products. Included are data and references translated from foreign sources, here appearing for the first time in U.S. scientific literature.

153 Milne, P. H. 1978. *Fish and Shellfish Farming in Coastal Waters*. Farnham, Surrey, England: Fishing News Books. 208 p.
This text covers site selection and design, construction, and culture methods for marine fish farming. The legal aspects of coastal zone use, the effects of fish

farming on the marine environment, and a useful bibliography are also included.

154 Moen, E., ed. 1984. *Cured Fish: Market Patterns and Prospects*. Rome: FAO. (FAO Fisheries Technical Paper, 233).

155 Morales, J. C. 1983. *Acuicultura Marina Animal*. Madrid: Ediciones Mundi-Prensa.
This Spanish-language text reviewed recent literature on mariculture.

156 Morel, F. M. 1983. *Principles of Aquatic Chemistry*. New York, NY: John Wiley. 446 p.

157 Munro, J. L. 1980. *Marine and Coastal Processes in the Pacific: Ecological Aspects of Coastal Zone Management*. Jakarta: UNESCO. Regional Office for Science and Technology.

158 Nriagu, J. O., ed. 1983. *Aquatic Toxicology*. New York: John Wiley. (Advances in Environmental Science and Technology series). 525 p.

159 *Ocean Science for the Year 2000*. 1984. Paris: UNESCO. 95 p.
This text examines projected trends in oceanographic research to the year 2000. The following topics are included: marine science, research methods, physical and chemical oceanography, marine biology, aquaculture, and the ocean floor.

160 Olausson, E. and **I. Cato**, eds. 1980. *Chemistry and Biochemistry of Estuaries*. New York, NY: John Wiley. 452 p.

161 Parker, S. P., ed. 1982. *Synopsis and Classification of Living Organisms*. New York, NY: McGraw-Hill. 2 vols.
Provides an excellent source of taxonomic information. Each entry describes a particular organism and its habitat, and cites key references for further reading.

162 Parsons, T. R., M. Takahashi, and **B. Harqarve.** 1984. *Biological Oceanographic Processes*. Oxford: Pergamon Press. 342 p.
This is an introductory text for students, oceanographers, engineers, hydrologists, and fisheries experts who require quantitative information on biological oceanographic processes.

163 Pillay, T. V. and **W. A. Dill**, eds. 1979. *Advances in Aquaculture*. Farnham, Surrey, England: Fishing News Books. 672 p.
Includes all the papers presented at FAO's Technical Conference in Aquaculture held in Kyoto, Japan, in 1976. A valuable text summarizing aquaculture development worldwide up to that year.

164 Piper, R. G., I. B. McElwain, L. E. Orme, *et al.* 1982. *Fish Hatchery Management*. Washington, DC: United States Department of the Interior, Fish and Wildlife Service. 517 p.
This text is an excellent reference in all aspects related to tropical and hatchery

requirements and operations; broodstock, spawning, and egg handling, nutrition, and feeding; fish health management; and transportation of live fish. Information is included for both temperate and tropical freshwater fish species.

165 Pitcher, T. J. and **P. J. B. Hart.** 1982. *Fisheries Ecology*. Westport, CT: AVI Publishing. 414 p.
The adaptive physiology and behaviour of fish, nutrition and production, and the economics of fisheries in terms of both energy and finance are discussed. The treatment is worldwide and includes reference to both freshwater and marine fisheries.

166 Ragotzkie, R. A. 1983. *Man and the Marine Environment*. Boca Raton, FL: CRC Press. 200 p.
An examination of the various types of interaction between man and the ocean environment. Has chapters on coastal law, marine recreation, the fishing communities, including the use of aquaculture, and undersea exploration.

167 Raymont, J. E. G. 1980. *Plankton and Productivity in the Oceans*. Oxford: Pergamon Press.
Vol. 1 *Phytoplankton*. 503 p.
Vol. 2 *Zooplankton*. 892 p.

168 Reay, P. J. 1979. *Aquaculture*. London: Edward Arnold; Baltimore, MD: University Park Press. (The Institute of Biology's Studies in Biology, 106). 64 p.
This is an account of aquaculture presented in terms of the modification of biological systems. Includes information on the biology of aquatic organisms and the process of domestication, as well as commercial applications.

169 Richardson, J. C., ed. 1985. *Managing the Ocean*. Mt. Airy, MD: Lomond Publications. 407 p.
Papers discussing science, law, economics, and engineering when applied to the ocean and coastal environment.

170 Riley, J. P. and **R. Chester.** 1971. *Introduction to Marine Chemistry*. London: Academic Press. 465 p.
Provides the basic background information on marine chemistry. It includes detailed chapters on salinity, physical processes, elements of seawater, dissolved gases, micronutrients, organic compounds, production, sedimentology of deep-sea sediments, and geochemistry.

171 Ross, D. A. 1979. *Opportunities and Uses of the Ocean*. Berlin: Springer-Verlag. 320 p.
Discusses shipping, marine resources, aquaculture, pollution, and marine law.

172 Rothschild, B. J. 1983. *Global Fisheries: Perspectives for the 1980s*. Berlin: Springer-Verlag. 289 p.
A collection of papers addressing issues of current importance in

fisheries management, for example Law of the Sea, fisheries research programmes, etc.

173 Ryther, J. H. and **J. E. Bardach.** 1968. *The Status and Potential of Aquaculture, Particularly Invertebrate and Algae Culture.* Springfield, VA: National Technical Information Service.

This book presents an examination of the present situation of aquaculture and its relevance to eliminating world hunger.

174 Schenck, H., Jr., ed. 1975. *Introduction to Ocean Engineering.* New York: McGraw-Hill. 351 p.

This text was designed for undergraduate and graduate students in engineering. It provides an introduction to the field of ocean engineering and examines hydrodynamics in coastal areas, sub-marine soil mechanics, ocean corrosion, selection of materials for ocean application, and thermal and structural limitations of ocean systems.

175 Shang, Y. C. 1981. *Aquaculture Economics: Basic Concepts and Methods of Analysis.* Boulder, CO: Westview Press. (Westview Special Studies in Agriculture/Aquaculture). 140 p.

This book examines the basic economics of developing and managing fish stocks in a variety of controlled environments. After outlining the basic economics of aquaculture and summarizing the applicable principles of production economics, the author introduces systems for fish-farm record keeping and data collection, and suggests methods for effective data analysis.

176 Silvester, R. 1974. *Coastal Engineering.* Amsterdam: Elsevier Science Publishers. (Developments in Geotechnical Engineering, 4A & B). 2 vols.

Vol 1 *Generation, Propagation and Influence of Waves.*

Vol. 2 *Sedimentation, Estuaries, Tides, Effluents, and Modelling.*

Volume 1 provides excellent descriptions and present methodology for the processing engineer in the fields of wave-generation processes, wave forecasting, shoaling water, wave recording, effect of structures, and effect of waves on structures. Volume 2 focuses on shoreline processes, longshore drift, coastal defence, long-period waves, estuarine problems, marine hydraulics and hydraulic modelling. These volumes evolved from ocean engineering courses developed at the Asian Institute of Technology, Bangkok.

177 Simmonds, K. R., ed. 1983. *U.N. Convention on the Law of the Sea.* Dobbs Ferry, NY: Oceana Publications. (New Direction in the Law of the Sea). 312 p.

178 Smith, L. J. and **S. Peterson.** 1982. *Aquaculture Development in Less Developed Countries: Social, Economic and Political Problems.* Boulder, CO: Westview Press. (Special Studies in Agricultural–Aquaculture Science). 175 p.

This book examines the factors that can determine the success or failure of aquaculture projects in developing countries, particularly in Latin America, Africa, and the Middle East. The authors present examples of social, economic, and political constraints on aquaculture development, consider the acceptability of aquaculture products for consumption, look at income, production, and technology allocation, contrast fishing and aquaculture activities, and discuss appropriate organization of aquaculture enterprises.

179 Stowe, K. 1983. *Ocean Science*. New York: John Wiley. 610 p.
An introductory college text for oceanic sciences. Subjects include the history of oceanography, the geoscience aspects of oceanography, chemical and physical oceanography, marine life, and ocean resources.

180 Valiela, I. 1984. *Marine Ecological Processes*. Berlin: Springer-Verlag. 536 p.
This text presents a discussion of marine ecosystems, communities, and populations and of how general ecological principles, derived from both terrestrial and freshwater systems, can be applied to marine ecosystems.

181 Walne, P. P. 1974. *Culture of Bivalve Molluscs: 50 Years' Experience at Conwy*. Farnham, Surrey, England: Fishing News Books. 189 p.
This text focuses on the culture of the European oyster, *Ostrea edulis* and *Crassostrea gigas*.

182 Waugh, G. 1983. *Aquaculture: Fisheries Management: Theoretical Developments and Contemporary Appendices*. Boulder, CO: Westview Press. 225 p.
This text presents management procedures based on bioeconomic modelling that integrate the population dynamics of fish revenues with the economic processes of harvesting and marketing. To illustrate the ability of bioeconomic modelling two case studies based on Australian fisheries are discussed.

183 Wheaton, F. W. 1977. *Aquacultural Engineering*. New York: John Wiley. 708 p.
This introduction to aquacultural engineering is divided into two parts. Part I concentrates on the interaction of the environment with aquatic organisms, while Part II emphasizes the engineering consideration of cultured, as opposed to fisheries, aspects of aquaculture.

184 Wickstead, J. H. 1976. *Marine Zooplankton*. London: Edward Arnold; Baltimore, MD: University Park Press. (The Institute of Biology's Studies in Biology, 62). 64 p.
This text provides a conceptually organized survey of zooplankton including history of zooplankton research, environment of the zooplankton, zoological composition, zooplankton in the ecology of the oceans, adaptation to the environment, and zooplankton as food for man.

6.5 Bibliographies

Bibliographies are collections of references, usually on a specific subject, which provide the user with an instant starting-point for research. They are generally more useful for retrospective rather than current searching. Bibliographies of bibliographies provide a good general source of information, for example, for those setting up a research collection.

FAO is the major agency in the field and has been particularly instrumental in compiling and distributing bibliographies in aquaculture-related fields.

185 *Annotated Bibliography of Textbooks and Reference Materials in Marine Sciences.* 1975. Paris: UNESCO, Intergovernmental Oceanographic Commission. (IOC Technical Series. Provisional Edition).
This publication lists over 1,000 references under the headings of oceanography, fisheries, geochemistry, geology, geographics, ocean engineering and technology, pollution review books, and popular texts. It is an excellent source for materials published prior to 1975.

186 **Coche, A. G.**, comp. 1982. *Aquaculture in Marine Waters: A List of Reference Books, 1961–1981.* Rome: FAO. (FAO Fisheries Circular, 723. Rev. 1). 18 p.
This bibliography lists 145 selected books providing information related to aquaculture in brackish water and seawater. Author, geographic, taxonomic, and subject indexes provide further assistance in locating the information required.

187 **Coche, A. G.**, comp. 1982. *List of FAO Publications Related to Aquaculture, 1966–1982.* Rome: FAO. (FAO Fisheries Circular, 744). 21 p.
This bibliography lists 158 selected FAO documents published during the period January 1966 to April 1982 and related to aquaculture. Author, geographic, taxonomic, and subject indexes are provided.

188 **Coche, A. G.**, comp. 1983. *Lists of Serials, Newsletters, Bibliographies and Meeting Proceedings Related to Aquaculture.* Rome: FAO. (FAO Fisheries Circular, 758).
This is the most comprehensive and current list of aquaculture information sources available. It contains 121 titles of regular serials, 87 titles of irregular serials, 44 titles of conferences, 15 references for proceedings, and 141 references for meetings held from 1966 to 1982. Author, geographic, and subject indexes are provided.

189 **Encarnacion, E. N.**, comp. 1982. *Aquaculture Economics Abstracts.* 1982. Laguna, The Philippines: Southeast Asian Regional Center for Graduate Study Research in Agriculture.
This text provides an annotated bibliography of aquaculture information on various species, with sections on marketing, socioeconomics and financing.

190 *FAO Documentation Current Bibliography.* 1976–. Rome: FAO.
This is a regularly updated index to FAO publications and library accessions. Subject and geographic indexes are included.

191 *FAO Documentation Fisheries 1983–1985.* 1986. Rome: FAO. Library and Documentation Systems Division, David Lubin Memorial Library. 117 p.
This bibliography contains details of documents and publications of the Fisheries Department of the Food and Agriculture Organization of the United Nations.

192 *FAO Fisheries Department List of Publications and Documents 1948–1978.* 1979. Rome: FAO. (FAO Fisheries Circular, 100. Rev. 3). 74 p.
This publication provides a listing of the publications, working papers, and miscellaneous documents published by the FAO Fisheries Department during the years 1948–78.

193 *FAO Library Catalogue of Monographs.* 1976–. Rome: FAO.
A set of microfiche indexes to the holdings of the FAO library from 1976 to date.

194 **Gilbert, H.** 1984. *Quick Bibliography Series: Aquaculture 1978–1984.* Beltsville, MD: U.S. Department of Agriculture. National Agriculture Library.
Includes 337 citations based on a search of AGRICOLA.

195 **Hillard, A.** and **S. Jhaveri.** 1981. *Fish Preservation: An Annotated Bibliography.* Narragansett Bay, RI: University of Rhode Island. Marine Advisory Service. (Technical Report series, 82). 54 p.

196 **Hougart, B.** 1985. *Practical Aquaculture Literature: A Bibliography.* Beltsville, MD: U.S. Department of Agriculture. (Bibliographies and Literature of Agriculture, No. 35). 40 p.
Provides a listing of 'how to' publications.

197 *List of Publications and Documents 1977–1984.* 1984. Rome: FAO. (FAO Fisheries Circular, 100. Rev. 3, Suppl. 1).
This is an excellent list of the publications published by the FAO Fisheries Department from 1977 to the first quarter of 1984.

198 *Postharvest Food Losses in Developing Countries: A Bibliography.* 1978. Washington, DC: National Research Council, Board on Science and Technology for International Development, and Agency for International Development, Office of Agriculture.

199 **Sarria, J. P.**, ed. 1982. *Fisheries Bibliography: A Bibliography of Southeast Asian Fisheries Literature.* Los Banos, The Philippines: Southeast Asian Regional Center for Graduate Study and Research in Agriculture.
This is the first major fisheries bibliography to be published by the Southeast Asian Regional Center for Graduate Study and Research in Agriculture (SEARCA) through its Agricultural Information Bank for Asia (AIBA) project. The major categories include fisheries, aquaculture, aquatic plants, limnology, and oceanography.

200 *Selected Bibliography on the Economic Aspects of Aquaculture, 1969 to 1979.* 1979. Rome: FAO. (FAO Fisheries Circular, 702. Rev. 1).
A revision and expansion of an earlier edition, this publication includes references of interest to aquaculturalists. It includes a list of key journals in the field.

201 **Wiktov, C. L.** and **L. A. Foster.** 1980–. *Marine Affairs Bibliography.* Halifax, Nova Scotia: Dalhousie Law School.

This bibliography is published quarterly and cumulated annually. It includes citations under three general headings: General Works, Law of the Sea, and Maritime Transportation and Communication. It includes author, corporate, and geographical indexes. The bibliography can be searched by author, title, subject, geographical area, or journal title. Searches of the corresponding database can be arranged by contacting the publisher.

6.6 Primary and review journals

Most current scientific literature in aquaculture is published in primary journals or other periodical literature rather than in books. Over seventy of the more important journals that are included in this *Keyguide* are covered in the printed secondary literature (e.g. *Aquatic Sciences and Fisheries Abstracts*) and in commercially available databases (see Section 8.2). A good way to determine which periodicals are available is to review published serials lists.

A common problem in determining what journals have been cited is to recognize their titles when these are given in abbreviated form. In this section the full serial titles are given. The *International List of Periodical Titles World Abbreviations* is published regularly by the International Centre for the Registration of Serials in Paris and should be used to determine the recognized abbreviation. Database and index publishers such as BIOSIS (*Biological Abstracts*) often also issue lists of the journals covered in their indices.

Selection Aids and Lists of Journal Titles

202 *Irregular Serials and Annuals: An International Directory 1987–88*. 1987. New York, NY: R. R. Bowker Company.
Designed as a companion to the well known *Ulrich's International Periodicals Directory* [206A], this publication is worldwide in scope and provides data on 34,000 serials, annuals, conference proceedings, and other publications issued irregularly or less frequently than twice a year.

203 *List of Serials Held in the FAO Fisheries Branch Library*. 1982. Rome: FAO. (FAO Fisheries Circular, 743).

204 *Lists of Serials, Newsletters, Bibliographies and Meeting Proceedings Related to Aquaculture*. 1983. Rome: FAO. (FAO Fisheries Circular, 758).
This useful publication cites key aquaculture publications, dates of their coverage, publisher's addresses, and title abbreviations. Subject, author, and institution indexes are included.

205 **Meadows, J.**, **D. Walker**, and **S. Barrich.** 1985. *Mussel: A Union List of Serials in Marine Science Libraries*. Gloucester Point, VA: Virginia Institute of Marine Science. (Special Scientific Report, No. 113). 19 pages, plus 21 microfiche.
To answer a need for resource sharing among marine science libraries, the

librarians at Virginia Institute of Marine Science compiled the serials holdings of fifty-five marine libraries in the United States, Canada, and Bermuda. Most participants are members of IAMSLIC, the International Association of Marine Science Libraries and Information Centers. *Mussel* contains approximately 10,000 titles and cross-references; each title is annotated with the symbol and holdings of one or more libraries. The serial titles and holdings are contained on a set of 21 microfiche. The accompanying printed matter includes a directory of the participating libraries and a guide to their interlibrary loan policies.

206 *Serial Sources for the BIOSIS Data Base*. Annual. Philadelphia, PA: BIOSIS. 353 p.
This annual volume includes the CODEN code and frequency of publication for the more than 22,000 journals scanned for the BIOSIS database.

206A *Ulrich's International Periodicals Directory 1987–88*. 26th ed. 1987. New York, NY: R. R. Bowker Company. 2 vols.
Lists some 70,800 periodicals in 542 subject areas, including one entitled 'Fish and fisheries'.

207 *World List of Aquatic Sciences and Fisheries Serial Titles*. 1975–. Rome: FAO. (FAO Fisheries Technical Paper, 147, plus Supplements 1–5).
A bibliography of the major marine science journals, with supplements to 1981. Research reports, conferences, and government publications are also included.

Primary and Review Journals with a Regional or International Focus

Most periodicals cover one or a number of specific subject areas that have either an international or regional focus.

FAO's *Marine Science Content Tables* (*MSCT*) and *Freshwater and Aquaculture Contents Tables* (*FACT*) offer a 'table of contents' service and can be scanned to assess the current contents of major journals. There are numerous primary journals that include papers relevant to the multidisciplinary field of aquaculture. They focus on such topics as aquaculture itself, toxicology, oceanography, engineering, estuaries, fisheries, cell biology, chemistry, pollution, management, technology, and fishing techniques.

208 *Advances in Marine Biology*. 1963–. Irregular. En. London: Academic Press.

209 *AFZ International Fishing Information*. Monthly. En. Butzbach, Federal Republic of Germany: MC Moderne.
This publication includes articles on fisheries worldwide as well as aquaculture, fish processing, and fisheries technology.

210 *Aquacultural Engineering: An International Journal*. 1981–. Quarterly. En. Barking, Essex, England: Elsevier Applied Science Publishers.
This journal takes an interdisciplinary approach to aquaculture, including papers on biological, engineering, chemical, and socioeconomic aspects of aquaculture.

211 *Aquaculture*. 1972–. Monthly. En. Amsterdam: Elsevier Scientific.
Includes international research papers on the explanation, improvement, and management of all aquatic food resources.

212 *Aquaculture Digest*. 1976–. Monthly. En. San Diego, CA: Bob Rosenberry (editor/publisher). (9434 Keamy Mesa Road, San Diego, CA 92126, U.S.A.)
This journal provides a monthly report on marine fish and shellfish farming. It focuses on salmon, oyster, and shrimp farming in the United States and also covers international developments, research, and a wide range of other aquaculture-related topics.

213 *Aquaculture Magazine*. 1974–. Bimonthly. En. Asheville, NC: Achill River Corporation.
This journal covers research on fisheries and aquaculture worldwide.

214 *Aquatic Botany*. 1975–. Monthly. En. Amsterdam: Elsevier Science Publishers.
This international scientific journal includes papers dealing with applied and fundamental research on submerged, floating, and emergent plants in marine and freshwater ecosystems.

215 *Aquatic Toxicology*. 1981–. Quarterly. En. Amsterdam: Elsevier Science Publishers.
Included in this journal are scientific papers dealing with the mechanisms of toxicity in aquatic environments and the understanding of responses to toxic agents at community, species, tissue, cellular and subcellular levels. Topics discussed included: aspects of uptake, metabolism and excretion of toxicants; toxicant-induced alterations in organisms; the development of procedures and techniques that significantly advance the understanding of processes; and events that produce toxic effects.

216 *Archiv für Hydrobiologie*. 1906–. Monthly. De, En, Fr. Stuttgart: International Association of Theoretical and Applied Limnology.

217 *Asian Aquaculture*. Irregular. En. Manila: SEAFDEC.
This publication includes highlights of aquaculture developments in Asia.

218 *Biological Bulletin*. 1898–. Bimonthly. En. Woods Hole, MA: Marine Biological Laboratory. Woods Hole Institute of Oceanography.

219 *Biological Oceanography Journal*. 1981–. Quarterly. En. New York, NY: Crane, Russak.
The contents of this journal focus on ecological oceanography with particular reference to primary and secondary production, biogeography, natural history, community structure, physiology, evaluation and systematics, and the relationship of the oceanic biota to the physical milieu.

220 *Bulletin of Marine Science*. 1951–. Bimonthly. En. Coral Gables, FL: University of Miami, Marine Laboratory.

221 *Canadian Journal of Fisheries and Aquatic Sciences*. 1901–. Monthly. En. Ottawa: Fisheries and Oceans Canada. Scientific Information and Publications Branch.

222 *Coastal Engineering*. 1977–. Quarterly. En. Amsterdam: Elsevier Science Publishers.

223 *Crustaceana*. 1960–. Bimonthly. En, Fr, De. Leiden: E. J. Brill.

224 *Ecology*. 1920–. Bimonthly. En. Brooklyn, NY: Ecological Society of America.

225 *Environmental Biology of Fishes*. 1976–. Quarterly. En. The Hague: Dr. W. Junk Publishers.
The contents of this journal include monographs, papers, and short notes on fisheries as related to their morphology, behaviour, physiology, ecology, populations as well as systems modelling, evolution, and succession.

226 *Estuaries*. 1978–. Quarterly. En. Lawrence, KS: Allen Press.
This is the journal of the Estuarine Research Federation. Its contents focus on research in any aspect of natural science applied to estuaries including management concerns and public policy.

227 *Fish Farming International*. 1973–. Monthly. En. London: Arthur J. Heighway Publications.

228 *Fisheries Research*. 1981–. Quarterly. En. Amsterdam: Elsevier Science Publishers.
This multidisciplinary journal focuses on papers in the areas of fishing technology, fisheries science, and fisheries management. It is intended for fisheries biologists, aquaculturalists, gear technologists, naval architects, fisheries economists, administrators, and policy-makers and legislators.

229 *Fishery Bulletin*. 1881–. Quarterly. En. Seattle, WA: U.S. Department of Commerce. National Oceanic and Atmospheric Administration. National Marine Fisheries Service.
This bulletin presents research reports and technical notes on investigations in fishery science, engineering, and economics. Approximately 17 percent of the coverage focuses on research undertaken in the tropics.

230 *Journal of Applied Ichthyology*. 1985–. Quarterly. En. New York, NY: Paul Parey Scientific Publishers.
Articles in this journal focus on ichthyology, aquaculture, marine fishes, toxicology, immunology, and fishing technology.

231 *Journal du Conseil*. 1926–. 3 per year. En, Fr, De. Copenhagen: Conseil International pour l'Exploration de la Mer.

This journal publishes original papers, notes, and letters to the editor, together with reviews in the broad subject field of marine and fisheries science especially in connection with living resources and their environment.

232 *Journal of Biological Chemistry*. 1905–. Monthly. En. Bethesda, MD: American Society of Biological Chemists.

233 *Journal of Experimental Marine Biology and Ecology*. 1967–. Monthly. En. Amsterdam: Elsevier Science Publishers.
Papers on all aspects of marine ecology are included, with particular emphasis on physiology, biochemistry, and behaviour of organisms in relation to their environment.

234 *Journal of Fish Biology*. 1969–. Monthly. En. London: Academic Press.
An international journal covering marine and freshwater biology.

235 *Journal of Fish Diseases*. 1978–. Bimonthly. En. Oxford: Blackwell Scientific.

236 *Journal of Ichthyology*. 1960–. Bimonthly. En. Silverspring, MD: Scripta Publishing Co.
This is an English-language translation of the U.S.S.R. Academy of Sciences journal *Voprosy Ikhtiologii*. Articles cover freshwater and marine fisheries information and allow the scientist to keep up to date with Soviet fisheries research.

237 *Journal of Marine Research*. 1937–. Quarterly. En. New Haven, CT: Sears Foundation for Marine Research, Kline Geology Lab.

238 *Journal of Physical Oceanography*. 1971–. Monthly. En. Boston, MA: American Meteorological Society.

239 *Journal of Physiology*. 1978–. Monthly. En. Cambridge, England: Cambridge University Press.

240 *Journal of the Marine Biological Association of the United Kingdom*. 1887–. Quarterly. En. Cambridge, England: Cambridge University Press.
Included in this journal are original articles and notes on all aspects of marine biology, including the biology of marine animals and plants, as well as work on relevant chemical, physical, hydrographic, and geophysical topics. Papers are also published on the rapidly developing techniques employed at sea for sampling, recording, capture, and observation of marine organisms, and chemical analysis of seawater.

241 *Journal of the World Aquaculture Society*. Quarterly. En. Baton Rouge, LA: Louisiana State University. World Aquaculture Society.
This journal contains contributed articles focusing on all aspects of aquaculture development.

242 *Limnology and Oceanography*. 1956–. En. Bimonthly. Lawrence, KS: Allen Press.

243 *Marine Behaviour and Physiology*. 1972–. En. 8 per year. London: Gordon & Breach.

244 *Marine Biology*. 1967–. 18 per year. En. New York, NY: Springer-Verlag New York Inc.
This is an international journal publishing research on plant and animal life in marine and coastal waters.

245 *Marine Chemistry*. 1972–. Bimonthly. En. Amsterdam: Elsevier Science Publishers.
An international journal that includes the results of studies of all chemical aspects of the marine environment.

246 *Marine Environmental Research*. 1978–. Monthly. En. Barking, Essex, England: Elsevier Applied Science Publishers.

247 *Marine Policy*. 1977–. Quarterly. En. Guildford, Surrey, England: Butterworth Scientific.
This journal includes papers and notes, and information on work in progress, meetings, and publications. The focus is on policies relating to the use of the oceans.

248 *Marine Pollution Bulletin*. 1970–. Monthly. En. New York, NY: Pergamon Press.
This journal includes news, comments, reviews, and research reports on marine pollution, management, and productivity.

249 *Marine Resource Economics*. 1984–. Quarterly. En. New York, NY: Crane Russak.
This journal includes articles that focus on economic development and management of fisheries.

250 *Ocean Engineering*. 1968–. Bimonthly. En. New York, NY: Pergamon Press.
This journal carries articles on design and construction of structures (including ships), sub-marine soil mechanics, coastal engineering, stress analysis, underwater instrumentation, aquacultural engineering, and underwater acoustics.

251 *Oceanography and Marine Biology: An Annual Review*. 1963–. Annual. En. Until 1974, London: George Allen & Unwin. 1975–, Aberdeen: Aberdeen University Press.

252 *Oceanologica Acta*. 1978–. Quarterly. Fr, En, De. Montrouge, France: Oceanologica Acta, CDR—Centrale des Revues.
This journal publishes original results, short notes, and reviews of work conducted in all sectors of oceanography including marine, estuarine, and brackish-water systems.

253 *Oceanus*. 1952–. Quarterly. En. Woods Hole, MA: Woods Hole Oceanographic Institution.

Three issues each year of this quarterly journal are devoted to a particular subject in marine science; the fourth issue is a 'general' one with numerous articles on a variety of topics.

254 *La Pêche Maritime.* 1919–. Monthly. Paris: Editions Maritimes.
This journal includes articles, studies, and bibliographies which focus on fisheries, aquaculture, and fisheries technology around the world.

255 *PESCA.* 1960–. Bimonthly. Es, En. Lima: Publicaciones S.A.
This publication includes articles on the fisheries and fishing industry in Latin America, and Peru in particular.

256 *Revista Latinoamericana de Acuicultura.* 1979–. Semi-annual. Es. Mexico City: Dep. Pesca. Dirección General de Acuacultura.
This review includes articles on aquaculture and related topics in Latin America.

257 *Science.* 1980–. Weekly. En. Washington, DC: American Association for the Advancement of Science.

258 *SWIO Fisheries Bulletin/Bulletin des Pêches OISO.* Quarterly. En, Fr. Victoria, Mahe, Seychelles: FAO/UNESCO. Projet pour le Développement et l'Aménagement des Pêches dans l'Ocean Indien Sud-Occidental.
This bulletin describes the fisheries of the member states of the Southwest Indian Ocean. Included are: Comoros, France, Kenya, Madagascar, Mauritius, Mozambique, the Seychelles, Somalia, and Tanzania.

Primary and Review Journals with a National Focus

Many countries publish their own national journals on aquaculture or in related fields. National journals, particularly for developing countries, have been included in the following list. These journals are usually published in the national language. They provide a useful forum for research on particular regional or national issues.

Algeria

259 *Pelagos.* 1963–. Semi-annual. Fr. Alger-Bourse, Algeria: Bulletin du Centre de Recherches Océanographiques et des Pêches.

Argentina

260 *Acta Oceanographia Argentina.* 1977–. Semi-annual. Es (En summaries). Buenos Aires: Comité Nacional para la International Association for the Physical Sciences of the Ocean.

261 *Centro de Investigación de Biología Marina. Contribución Técnica.* 1965–. Irregular. Es. Buenos Aires: Centro de Investigación de Biología Marina.

262 *Revista de Investigación y Desarrollo Pesquero.* Es. Mar del Plata: Instituto Nacional de Investigaciones y Desarrollo Pesquero (INIDEP).

Australia

263 *Australian Fisheries*. 1942–. Monthly. En. Canberra: Australian Government Publishing Service.
This journal focuses on fisheries and fishing technology in Australia, New Zealand, and the Pacific.

264 *Australian Journal of Marine and Freshwater Research*. 1950–. Bimonthly. En. Melbourne: Commonwealth Scientific and Industrial Research Organization.

Brazil

265 *Boletim de Ciências do Mar*. (Title varies). 1961–. Irregular. Pt, En. Ceará, Brazil: Universidade Federal do Ceará. Laboratorio de Ciências do Mar.

266 *Boletim Informativo del CERLA*. Quarterly. Pt. São Paulo: CERLA.

Canada

267 *Bulletin*. 1986–. Quarterly. En, Fr. St. Andrews, New Brunswick: Aquaculture Association of Canada.
The *Bulletin* contains the proceedings of the annual meeting of the Association, short articles, news items, information on meetings, letters to the editor, interesting photographs and favourite recipes.

268 *Canadian Aquaculture Bulletin*. 1984–. Bimonthly. En. Victoria, British Columbia: Harrison House Publishers.
This bulletin contains articles on the developments of aquaculture in Canada and Canadian involvement in international aquaculture projects.

Chile

269 *Ciencia y Tecnología del Mar*. 1975–. Semi-annual. Es (En summaries). Valparaiso: Comité Oceanográfico Nacional.

270 *Investigación Pesquera*. Annual. Es (En summaries). Cable: Instituto de Fomento Pesquero. Santiago.

China

271 *Acta Oceanologica Sinica*. 1980–. 3 per year. En. Tientsin, China: Chinese Society of Oceanography. Marine Scientific and Technological Data Centre.

272 *China Fisheries Monthly*. En. Taipei, Taiwan (Republic of China): China Fisheries Association.

273 *Chinese Journal of Oceanology and Limnology*. 1982–. Semi-annual. En. Beijing: Science Press.
This journal publishes English translations of papers selected from Chinese journals as well as original contributions by leading scientists internationally.

274 *Journal of Fisheries of China*. En. Editorial Committee of the China Society of Fisheries. People's Republic of China.

275 *Marine Fisheries Research*. Ch (En summaries). Shandong, China: Yellow Sea Fisheries Research Institute.

276 *Oceanologia et Limnologia Sinica*. 1970–. Bimonthly. En. Qingdao, China: Chinese Society of Oceanology and Limnology.

277 *Studia Marina Sinica* (Hai Yang K'o Hsuch Chi K'an). 1963?–. Ch (En summaries). Tsingtao, China: Institute of Oceanology. Academia Sinica.

Colombia

278 *Anales del Instituto de Investigaciones Marinas de Punta de Betín*. Es, En. Santa Marta, Colombia: INVEMAR.

279 *Boletín Científico*. Es (En summaries). Cartagena, Colombia: Centro de Investigaciones, Oceanográficas Hidrográficas.

Costa Rica

280 *Revista de Biología Tropical*. 1953–. Semi-annual. Es, En. San José, Costa Rica: Universidad de Costa Rica.

Cuba

281 *Estudios*. 1966–. Es (En summaries). Havana: Instituto de Oceanología. Academia de Ciencias de Cuba.

Ecuador

282 *Aquanet Cultivos*. 1984–. Quarterly. Es. Guayaquil, Ecuador: Aquanet Cultivos.

This trade journal focuses on aquaculture operations in Ecuador and around the world.

283 *Boletín Científico y Técnico*. 1978–. Irregular. Es, Fr. Guayaquil, Ecuador: Instituto Nacional de Pesca.

Finland

284 *Finnish Fisheries Research*. 1972–. Irregular. En. Helsinki: Finnish Game and Fisheries Research Institute.

285 *Finnish Marine Research*. 1978–. Irreg. Helsinki: Merentutkimuslaitos (Institute of Marine Research).

France

286 *Cybium*. 1977–. Quarterly. Fr (En summaries). Paris: Muséum d'Histoire Naturelle. Société Française d'lchtyologie.

Cybium is the bulletin of the Ichthyological Society of France. It includes original articles, reviews, and bibliographies on freshwater and marine fishes.

287 *Tethys*. 1969–. Quarterly. Fr. Marseilles: Université Aix-Marseille. Station Marine d'Endoume.

288 *Vie et Milieu. Série A—Biologie Marine*. 1950–. Fr. Paris: Bulletin du Laboratoire Arago. Université de Paris.

German Democratic Republic

289 *Beiträge zur Meereskunde*. 1961–. De. Berlin: Akademie der Wissenschaften der D.D.R. Institut für Meereskunde.

India

290 *Fish Technology Newsletter*. Quarterly. En. Cochin, India: Central Institute of Fisheries Technology.

291 *Fishery Technology*. 1964–. 2 per year (Jan. and July). En. Cochin, India: Society of Fisheries Technologists.

292 *Indian Journal of Fisheries*. 1954–. 2 per year. En. Cochin, India: Indian Council for Agricultural Research.

293 *Indian Journal of Marine Sciences*. 1972–. Quarterly. En. New Delhi: Council of Scientific and Industrial Research.

294 *Journal of Marine Biological Association of India*. 1959–. En. 2 per year. Cochin, India: Marine Biological Association of India.

295 *Mahasagar. Bulletin of National Institute of Oceanography*. 1968–. Quarterly. En. Goa, India: National Institute of Oceanography.

Ireland

296 *Aquaculture Ireland*. En. Dun Laoghaire, Co. Dublin, Ireland: BIM, in cooperation with the Irish Aquaculture Association.
This magazine focuses on topics of interest to the aquaculture industry in Ireland.

Israel

297 *Bamidgeh: Bulletin of Fish Culture in Israel*. 1949–. Quarterly. En. Nir-David, Israel: Fish Breeders Association.

Japan

298 *Bulletin of the Japanese Society of Scientific Fisheries*. 1932–. Monthly. Ja, En. Tokyo: Tokyo University of Fisheries.

299 *Bulletin of the National Research Institute of Aquaculture*. 1980–. Semi-annual. Ja, En. Japan: National Research Institute of Aquaculture.

300 *Journal of the Oceanographical Society of Japan*. 1942–. Bimonthly. En, Ja. Tokyo: Oceanographical Society of Japan.

Korea

301 *Bulletin of Korea Ocean Research and Development Institute*. 1979–. En, Ko. Seoul: Korea Ocean Research and Development Institute.

302 *Journal of the Oceanological Society of Korea*. 1966–. En, Ko. Seoul: Oceanological Society of Korea. Dept. of Oceanography. College of Natural Sciences. Seoul National University.

Kuwait

303 *Kuwait Bulletin of Marine Science*. 1980–. Irregular. En. Salmiya, Kuwait: Kuwait Institute for Scientific Research.
An irregular series of bulletins on subjects related to marine science in the Arabian Gulf and the Gulf of Oman.

Libya

304 *Bulletin of the Marine Research Centre*. En. Tripoli: Marine Biology Research Center.

Mexico

305 *Ciencias Marinas*. 1974–. Es (En summaries). Ensenada, Mexico: Universidad Autónoma de Baja California. Unidad de Ciencias Marinas.
This journal includes scientific papers related to the development of marine resources, particularly in Mexico.

306 *Pesca y Marina*. 1948–. Bimonthly. Es. Fijuome, BC, Mexico: Editorial Pesca y Marina S.A., Delegación Mexicana.

307 *Tecnica Pesquera: La Revista de la Pesca Mexicana*. 1967–. Monthly. Es. Mexico City: Ediciones Mundo Marino.

New Zealand

308 *New Zealand Journal of Marine and Freshwater Research*. 1967–. Quarterly. En. Wellington: Department of Scientific and Industrial Research.

309 *Shellfisheries Newsletter*. Quarterly. En. Wellington: Fisheries Research Division.

Norway

310 *Sarsia*. 1961–. Irreg. Universitetet i Bergen, Norway.

Philippines

311 *Fish Marketing Review*. Quarterly. En. Quezon City, The Philippines: Fisheries Development Authority.

Spain

312 *Investigación Pesquera.* 1955–. Irregular. Es. Barcelona: Instituto de Investigaciones Pesqueras.

Tunisia

313 *Bulletin de l'Institut National Scientifique et Technique d'Océanographie et de Pêche de Salammbô.* 1966–. Fr. Salammbô, Tunisia: Institut National Scientifique et Technique d'Océanographie et de Pêche de Salammbô.

Union of Soviet Socialist Republics

314 *Oceanology.* 1965–. Bimonthly. En. Washington, DC: American Geophysical Union.
Describes research being undertaken by scientists in the U.S.S.R. in the fields of marine physics, marine chemistry, marine geology, and marine biology. This is a translation of the Russian-language journal *Okeanologiia* [315].

315 *Okeanologiia.* 1961–. Bimonthly. Ru. Moscow: Akademiia Nauk SSR.
This is the original version of the English translation *Oceanology* [314].

United Kingdom

316 *Fisheries Management.* 1970–. Quarterly. En. Oxford: Blackwell Scientific.

United States of America

317 *North American Journal of Fisheries Management.* 1980–. Quarterly. En. Bethesda, MD: American Fisheries Society.

318 *Progressive Fish Culturist.* 1938. Quarterly. En. Bethesda, MD: American Fisheries Society.

319 *Transactions of the American Fisheries Society.* 1972–. Bimonthly. En. Bethesda, MD: American Fisheries Society.

Venezuela

320 *Boletín del Instituto Oceanográfico, Universidad de Oriente, Cumaná.* 1961–. Irregular. Es (En summaries). Cumaná, Venezuela: Universidad de Oriente. Instituto Oceanográfico.

321 *Cuadernos Oceanográficos.* 1961–. Es. Cumaná, Venezuela: Universidad de Oriente. Instituto Oceanográfico.

6.7 Abstract journals and secondary sources

Secondary journals represent an excellent source of current and historical information. Many are not simply lists of citations but also contain indexes to subject, title, author, and geographic coverage.

The most comprehensive abstract journal for aquaculture-related topics is

Aquatic Sciences and Fisheries Abstracts, Part 1: *Biological Sciences and Living Resources*; Part 2: *Ocean Technology, Policy and Non-Living Resources*. It can be accessed in print form or through database searches (Section 8.2).

AGRINDEX (see the description of FAO near the start of Part III) provides a list of citations by subject areas. It includes a section on marine sciences and fisheries that includes aquaculture-related citations. It is valuable to librarians or information scientists but because it does not contain abstracts it does not provide enough information to be useful to the user who requires more information than just the author and title, and bibliographic citation. (*AGRINDEX/AGRIS* is now also available in database form from the DIALOG service.)

AGRIASIA (again, see the description of FAO in Part III) has the same format as *AGRINDEX* but focuses on agriculture information (including marine sciences, fisheries, aquaculture) from Asia. It is particularly useful to librarians and information scientists either located in Asia or requiring information relating to Asia.

322 *AGRIASIA*. Quarterly. En, (language of original document). Laguna, The Philippines: AIBA/SEARCA.
AGRIASIA includes fisheries and aquatic sciences citations.

323 *Aquaculture Abstracts*. 1984–. Quarterly. En. Bethesda, MD: Cambridge Scientific Abstracts.
A quarterly compilation of all the citations related to aquaculture that have appeared in *Aquatic Sciences and Fisheries Abstracts*.

324 *Aquatic Sciences and Fisheries Abstracts*. 1969–. Monthly. En. Bethesda, MD: Cambridge Scientific Abstracts.
Part 1 *Biological Sciences and Living Resources*.
Part 2 *Ocean Technology, Policy and Non-Living Resources*.
This abstracting journal, published as a cooperative effort of FAO, IOC, and national marine science organizations, includes annotated citations in twenty-seven major subject categories from aquaculture to geology, oceanography, and water pollution. Over 5,000 primary journals as well as books, monographs, proceedings, technical reports are monitored for inclusion.

325 *Biological Abstracts*. 1927–. Semi-monthly. En. Philadelphia, PA: BIOSIS.
Citations are selected from the nearly 8,000 primary journal and monograph titles that are monitored. Some of the subjects included which relate to marine sciences are bacteriology, bioengineering, genetics, microbiology, nutrition, parasitology, systematic biology, toxicology, and zoology.

326 *CAB Abstracts*. 1973–. Monthly. En. Slough, England: Commonwealth Agricultural Bureaux.
Contains abstracts from the thirty-one journals published by the Commonwealth Agricultural Bureaux. Subjects areas covered include breeding, disease, nutrition, and aquaculture.

327 *Ecology Abstracts*. 1975–. Monthly. En. Bethesda, MD: Cambridge Scientific Abstracts.
This journal covers research on the interactions between organisms and their environment, including ecosystems in land, freshwater, and marine environments.

328 *Environmental Abstracts*. 1971–. Monthly. En. New York, NY: Environmental Information Center.
The major subject areas covered which are relevant to marine sciences are: oceans and estuaries, renewable resources (water), and water pollution.

329 *Food Science and Technology Abstracts*. 1969–. Monthly. En. Frankfurt am Main: International Food Service.
Relevant subjects covered include: food engineering, food packaging, and fish and food laws.

330 *Freshwater and Aquaculture Contents Tables* (*FACT*). 1978–. Monthly. En, Fr, Es, Ru, De. Rome: FAO.
This free serial reprints the tables of contents of major journals in the fields of aquaculture and freshwater biology. It is an excellent tool for current awareness, particularly for libraries that cannot afford expensive abstracting journals.

331 *Marine Science Contents Tables*. 1966–. Monthly. En, Fr, Es, Ru, De. Rome: FAO.
An excellent current-awareness reference for all libraries and scientists. This free periodical compiles the tables of contents for the main marine science journals.

332 *Pollution Abstracts*. 1970–. Bimonthly. Bethesda, MD: Cambridge Scientific Abstracts.
This abstracting journal is a good resource for environmentally related technical literature or publications. Subjects relevant to marine sciences include marine pollution and aquatic toxicology.

6.8 Handbooks and manuals

Handbooks and manuals are particularly aimed at the aquaculturalist, practising scientist, oceanographer, fisherman, fish processor, researcher, technician, etc. Methods manuals such as *Standard Methods for the Examination of Water and Waste Water* [366], laboratory manuals such as Anderson's (1979) *Introductory Oceanography Laboratory Manual* [333], and material selection handbooks such as FAO's Simple Methods for Aquaculture series [336, 337] are a few of the many subject-specific handbooks and manuals available.

FAO and the Network of Aquaculture Centres in Asia (NACA), a regional project of FAO, have published many aquaculture handbooks, manuals, and 'how to' books that offer excellent instructions and information, particularly for individuals working in aquaculture in developing countries.

CRC Press has also published subject-specific handbooks such as those of

Geyer (1983), Komar (1983) [348], McVey (1983) [356] and Moore *et al.* (1974–) [361] which are excellent reference guides for the collections of information centers, libraries, and the personal libraries of researchers.

333 **Anderson, E. E.** 1979. *Introductory Oceanography Laboratory Manual.* Minneapolis, MN: Burgess Publishing.
A handbook of tables for oceanographic data compilation, for measuring salinity, oxygen solubility, temperature, etc.

334 **Baker, J. T.** and **V. Murphy.** 1976–. *Compounds from Marine Organisms.* Cleveland, OH: CRC Press. (CRC Handbook of Marine Science). 2 vols.

335 **Bialek, E. L.**, comp. 1966. *Handbook of Oceanographic Tables.* Washington, DC: U.S. Naval Oceanographic Office. (Special Publication, 68).
A volume of reference tables for the oceanographer and engineer. The first section gives physical oceanic data (for example, temperatures of the oceans), while the second part consists of tables for computation and conversions.

336 **Coche, A. G.** 1985. *Soil and Freshwater Fish Culture.* (FAO Training Series, *Simple Methods for Aquaculture*, Vol. 6). Rome: FAO. 174 p.
This manual provides instruction on how to plan and make a soil survey, soil properties, chemical properties, soil suitability, and soils and freshwater fish culture.

337 **Coche, A. G.** and **H. Vanderwal.** 1981. *Water for Freshwater Fish Culture.* (FAO Training Series, Simple Methods for Aquaculture, Vol. 4). Rome: FAO. 111 p.
This family basic manual highlights water requirements, water flow, and water storage.

338 **de la Cruz, C. R.** *Fishpond Engineering: A Technical Manual for Small- and Medium-Scale Coastal Fish Farms in Southeast Asia.* Manila: FAO/UNDP South China Sea Fisheries Development and Coordinating Program, 1983. (SCS Manual, No. 5). 180 p.

339 **Dexter, S. C.** 1979. *Handbook of Oceanographic Engineering Materials.* New York, NY: John Wiley.
This handbook will aid in materials selection for oceanographic engineers, designers, planners, and researchers. It is limited to the metals, alloys, plastics, and other nonmetallic materials that are commonly in seawater environments.

340 **Gosner, K. L.** 1974. *Guide to Identification of Marine and Estuarine Invertebrates.* New York, NY: Wiley-Interscience. 693 p.
An introductory guide to the invertebrate fauna of the northeastern United States. Initial chapters describe the environment and the main body of the text describes each phylum. General and systematic indexes are included.

341 Grasshoff, K., M. Ehrhardt, and **K. Kremling**, eds. 1983. *Methods of Seawater Analysis*. Weinheim: Verlag Chemie. 419 p.

342 Grofit, E. 1980. *The Fishing Technology Unit: A Guide for the Planning, Establishment and Operations of National Fishing Technology Services in Developing Countries*. 1980. Rome: FAO. (FAO Fisheries Technical Paper, 199), 55 p.

343 Hamilton, L. S. and **S. C. Snedakes**, eds. 1984. *Handbooks for Mangrove Area Management*. Nairobi: United Nations Environment Programme; Honolulu, HI: East-West Center Environment and Policy Institute. 123 p.

344 *Handbook on Field Identification of Fishes, Crustaceans, Molluscs, Shells, and Important Aquatic Plants*. Manila: Bureau of Fisheries and Aquatic Resources (BFAR). FAO/UNDP South China Sea Fisheries Development and Coordinating Program, 1978. (SCS Manual, No. 2). 60 p.

This handbook is primarily designed for the field enumerators engaged in the collection of fishery data. It can also provide a common framework of names and identification for exchange of information between fishery biologists, statisticians, economists, and planners.

345 Head, P. C., ed. 1985. *Practical Estuarine Chemistry: A Handbook*. London: Cambridge University Press. (Estuarine and Brackish-Water Science Association Handbook). 350 p.

346 Jhingran, V. G. and **R. S. Pullin.** 1983. *A Hatchery Manual for the Common Chinese and Indian Major Carps*. Manila: Asian Development Bank. 191 p.

This manual discusses the biology of cultured carp, components of a carp hatchery, broodfish care, induced spawning, post-larvae and fry rearing, transplant, applied genetics, nutrition, diseases, and routine upkeep.

347 Kennedy, W. A. 1978. *A Handbook on Rearing Pan-Size Pacific Salmon Using Floating Seapens*. Fisheries and Marine Service Industry Report No. 107. Fisheries and Environment Canada. 111 p.

This manual details the needs to establish an aquaculture operation for rearing pan-size Pacific salmon, *Oncorhynchus* spp.

348 Komar, P. D., ed. 1983. *CRC Handbook of Coastal Processes and Erosion*. Boca Raton, FL: CRC Press. 320 p.

A discussion of the effects of rising seawater levels on the coastal zone environment. The focus is on erosion and its impacts on beaches and shoreline structures.

349 Kungvankij, P., T. E. Chua, *et al.* 1986. *Shrimp Culture: Pond Design, Operation and Management*. Bangkok: NACA. (NACA Training Manual Series, No. 2). 68 p.

This manual includes sections on: pond culture, site selection, species culture, pond design and construction, pond preparation, seed supply, culture techniques, water quality management, feeds and feeding, manipulation of stocking, and harvesting and preservation.

350 **Kungvankij, P., L. B. Tivo,** *et al.* 1986. *Shrimp Hatchery Design, Operation and Management.* Bangkok: NACA. (NACA Training Manual Series, No. 1). 88 p.
This manual provides information on the following aspects of shrimp culture: hatchery siting, design and construction; the life cycle of shrimp; spawning; larvae feed; egg collection, hatchery and transportation of nauplii, larvae rearing, hatchery management, and harvest and transport of larvae.

351 **Kungvankij, P., L. B. Tivo, Jr., B. J. Padadera, Jr., I. O. Potestus,** and **T. E. Chua.** 1986. *An Improved Traditional Shrimp Culture Technique for Increasing Pond Yield.* Bangkok: NACA. (NACA Technology Series). 14 p.
This manual includes sections on ponds, seed supply, pond preparation, pond management, harvesting, and financial analysis.

352 **Kungvankij, P., L. B. Tivo, Jr., B. J. Padadera, Jr.,** *et al.* 1986. *A Prototype Warmwater Shrimp Hatchery.* Bangkok: NACA. (NACA Technology Series). 32 p.
This manual includes sections on hatchery design, operation, and management; larvae rearing, harvesting, and transportation, and financial analysis.

353 **Kungvankij, P., L. B. Tivo, Jr., B. J. Pudadera, Jr.,** and **I. O. Potestus.** 1986. *Biology and Culture of Sea Bass* (Lates calcarifer). Bangkok: NACA. (NACA Training Manual Series, No. 3). 70 p.
This manual includes sections on biology, hatchery design, seed production and hatchery techniques, culture, and financial analysis of sea bass culture.

354 **Leitritz, E.** and **R. C. Lewis.** 1980. *Trout and Salmon Culture (Hatchery Methods).* Berkeley, CA: California Department of Fish and Game. (Fish Bulletin, No. 164). 197 p.
This manual includes discussion on hatchery water supply, spawning, picking eggs, fish eff incubators, trout and salmon ponds, feeding practices, transportation, diseases, treatment, and the classification of fishes.

355 **McLarney, Wm.** 1984. *The Freshwater Aquaculture Book.* Point Roberts, WA: Hartley & Marks. 583 p.
This is a handbook for small-scale fish culture in North America.

356 **McVey, J. P.,** ed. 1983. *CRC Handbook of Mariculture.* Boca Raton, FL: CRC Press.
A handbook of methods used in the culture of crustaceans, particularly freshwater prawn, penaeid shrimp, and lobster. A comprehensive bibliography is included.

357 *Manual of Fisheries Science.* 1979. Rome: FAO. (FAO Fisheries Technical Paper, 115. Rev.).

358 *Manual of Methods for Chemical Analysis of Water and Waste Water.* 1976. Washington, DC: Environmental Protection Agency.

359 *Manual of Sampling and Statistical Methods for Fisheries Biology*. 1966. Rome: FAO. (FAO Manuals in Fisheries Science, 3).
Part 1 *Sampling Methods*.
One of a series of manuals discussing statistical methods for fisheries biologists.
Part 2 *Methods of Resource Investigation and Their Application*.

360 *Manual on Pond Culture of Penaeid Shrimp*. 1978. Manila: ASEAN/FAO/UNDP South China Sea Fisheries Development and Coordinating Program. 132 p.
This manual covers all aspects of the culture of penaeid shrimp from suitability of various species to harvesting.

361 **Moore, J. R.** *et al.*, eds. 1974–. *CRC Handbook of Marine Science*. Boca Raton, FL: CRC Press.
A series of handbooks that discuss chemical and physical oceanography, marine biology, and marine products.

362 **New, M. B.** and **S. Singholka.** 1982. *Freshwater Prawn Farming: A Manual for the Culture of* Macrobrachiron rosenbergii. (FAO Fisheries Technical Paper, 225). 116 p.
This manual was prepared on the basis of experience in Thailand in the farming of the giant freshwater prawn, *Macrobrachiron rosenbergii*.

363 **Novikov, V. M.**, ed. 1983. *Handbook of Fishery Technology*. Rotterdam: Balkema. 504 p.

364 **Sorgeloos, P., P. Larens, P. Leger,** *et al.* 1986. *Manual for the Culture and Use of Brine Shrimp*, Artemica, *in Aquaculture*. Ghent: State University of Ghent. 319 p.
This manual includes sections on the biology and ecology of *Artemica*, cyst morphology and metabolism, harvesting, processing and storage of cysts, production of nauplii, use of metanauplii as a food source, production of *Artemica* in nature, and site selection and construction details for *Artemica* ponds.

365 **Sournia, A.** 1978. *Phytoplankton Manual*. Paris: UNESCO. (Monographs on Oceanographic Methodology, 6). 337 p.

366 *Standard Methods for the Examination of Water and Wastewater*. 1981. Washington, DC: American Public Health Association, American Water Works Association, and Water Pollution Control Federation. 1,200 p.
The first edition of this standard text was published in 1905. This is the basic handbook for any water analyses. The major chapters cover: physical examination, determination of metals, inorganic nonmetallic constituents, organic constituents, automated laboratory analyses, radioactivity analyses, bioassay methods, and microbiological and biological examination of water.

367 **Stevenson, J. P.** 1980. *Trout Farming Manual*. Farnham, Surrey, England: Fishing News Books. 183 p.
Covers all aspects of the culturing of trout.

368 Strickland, J. P. and T. R. Parsons. 1972 *A Practical Handbook of Seawater Analysis*. Ottawa: Fisheries and Oceans Canada. (Bulletin of the Fisheries Research Board of Canada, 167).

6.9 Statistical sources

369 *INFOFISH Trade Data Package*. Kuala Lumpur: FAO.
Available as a package of international market studies for fisheries products from Southeast Asia. The set includes volumes on the international markets for tuna, shrimp, cephalopods, fishmeal and seaweed, and regional markets for dried fish and finfish. Also included is a register of import regulations for fish and fishery products.

370 *INFOFISH Trade News*. Biweekly. En. Kuala Lumpur: FAO.
This fortnightly newsletter gives market trends and prices, commentary and forecasts for world fishery markets.

371 *Yearbook of Fishery Statistics*. Annual. En, Fr, Es. Rome: FAO. (FAO Fisheries series).
Each year two volumes are issued, entitled respectively *Catches and Landings* and *Fishery Commodities*. Detailed statistics, usually covering a four-year period, on fish catches and product shipments are given. Tables give access to the information by country/area, species/product, and import/export.

6.10 Audio-visual aids

Audio-visual aids are excellent materials to incorporate into training courses. Since video machines are now commonly found throughout the developing world, video tapes have become an inexpensive and effective way to disseminate aquaculture materials, particularly for extension purposes.

FAO's filmstrips provide excellent 'how to' information and have scripts translated into many languages.

372 **Emery, L.**, comp. 1982. *Aquaculture Audio-visual Films Catalog*. Bethesda, MD: American Fisheries Society.
This inexpensive catalogue lists films on aquaculture with a brief description of the subject of each, the supplier's address, and rental/purchase prices.

373 *FAO Filmstrips*. Rome: FAO.
This general catalogue includes listings for FAO fisheries filmstrips.

7 AQUACULTURE LIBRARIES

Libraries offer a wide array of resources and services to their users, according to the type of library, its role in the organization it serves, and the information needs of its patrons. Accessing and using library resources, however, requires an understanding of library procedures and organization, and in particular the use of the catalogue.

Even so, the main library catalogue should be regarded as only the starting point for finding information. Abstract journals to journal and monograph literature provide bibliographic citations to additional documents (also see Section 1.7). Many libraries also have the capacity for conducting literature searches of on-line bibliographic databases. The librarian will assist patrons in using the printed indexes, or can usually arrange for a database search (see Section 8.2 for more information on database searches).

Even if the library has extensive holdings, the use of indexes and bibliographic databases almost always retrieves references to items not available in its own collection. For this reason interlibrary loans are indispensable. An interlibrary loan can provide the user with a photocopy of a journal article or conference paper, or a duplicate of a microfiche document. Textbooks or theses can be obtained for loan on a short-term basis. Through interlibrary loans, library information resources are accessible worldwide; however, these services are often expensive and involve lengthy time lags for users outside of Europe and North America.

Libraries provide numerous other services to their users, and should not be overlooked as a source of information. Many libraries themselves issue a variety of publications including newsletters, bibliographies on special subjects, accessions lists of new library titles, serials lists showing journal holdings, and guides to using the library collection. Most libraries participate in library networks and can provide their users with addresses and key contacts for other institutions working in the area of interest. Also, librarians usually scan current periodicals for items of topical interest in order to provide a current awareness service. Libraries are also sources of information on upcoming conferences and workshops, university and college courses, and new publications. Every library's role in information retrieval and storage is different, and users should contact the librarian for a tour of their library and an explanation of the services and resources available to them.

7.1 Library directories (also see Part III)

374 Lengenfelder, H. 1983. *World Guide to Libraries*. Munich: K. G. Saur. (Handbook of International Documentation and Information, Vol. 8). 1,186 p.

An international directory to over 43,000 government, academic, public, and special libraries. Entries are arranged by country and by type of library.

375 Lengenfelder, H. 1983. *World Guide to Special Libraries*. Munich: K. G. Saur. (Handbook of International Documentation and Information, Vol. 17). 990 p.

Provides a subject index to over 32,000 special libraries. Each entry lists the address, collection statistics, and library services offered.

376 **Winn, C. P.** 1981. *Directory of Marine Science Libraries and Information Centers*. Woods Hole, MA: Woods Hole Oceanographic Institution, Sea Grant Program, and International Association of Marine Science Libraries and Information Centers. 72 p.

Aquaculture Information

377 **Barnes, F.** 1981. *Fish Technology Library Services in the IPFC Region*. Bangkok: FAO. Regional Office for Asia and the Pacific. (Indo-Pacific Fishery Commission Occasional Paper, 1981/1).

A discussion of fish technology information in the Indo-Pacific Fishery Commission region.

378 **Barnett, J. B.** 1984. 'Marine science libraries'. *Special Libraries* 75 (3): 183–192.

Examines the results of a survey of 156 marine science libraries. Library services, collections, staffing, and administration procedures are summarized.

379 **Edel, B. M.** and **J. B. Barnett.** 1981. 'Marine resources information: the National Sea Grant Depository'. *Special Libraries* 72: 59–62.

The National Sea Grant Program in the United States is one of the largest publishers of marine science information. This article describes the services available through the National Sea Grant Depository Library, including publications, literature searches, and reference services.

380 *Fisheries Information Science in Southeast Asia*. 1982. Bangkok, Thailand: Southeast Asian Fisheries Development Center and the International Development Research Centre.

The proceedings of the first seminar on fishery information science in Southeast Asia, held in Bangkok on 16–20 August 1982. It reviews existing fisheries information programmes in the region and includes proposals for improving the management of fisheries information in Southeast Asia.

381 **Grundy, R. L.** and **R. T. Ford**, eds. 1985. *Year of the Oceans: Science of Information Handling*. Woods Hole, MA: IAMSLIC. (International Association of Marine Science Libraries and Information Centers Conference series). 287 p.

The proceedings of the 10th IAMSLIC Conference held at Woods Hole, Massachusetts, in 1984. Library automation, networks, and organizations are discussed. These proceedings provide a useful and comprehensive survey of the state of marine science librarianship.

382 **Turnbull, D. A.** 1981. 'Fisheries and aquaculture information: the state of the art'. *Agricultural Information to Hasten Development. Proceedings of the 6th World Congress of the International Association of Agricultural Librarians and Documentalists, held 3–7 March, 1980, Manila, Philippines.* Laguna, The Philippines: Agricultural Libraries Association of the Philippines and Agricultural Information Bank for Asia.

Traces the development of fisheries and aquaculture industries worldwide and of the corresponding networks and documentation systems that have arisen to deal with information in these fields.

7.2 Library organization, cataloguing, and administration

383 *Anglo-American Cataloguing Rules*. 2nd ed. M. Gorman and P. W. Winkler, eds. 1978. Chicago, IL: American Library Association; London: Library Association. 400 p.
AACR2 is the cataloguing 'standard' for libraries.

384 **Anthony, L. J.**, ed. 1982. *Handbook of Special Librarianship and Information Work*. 5th ed. London: Aslib. 416 p.

385 **Fisher, E. L.** 1966. *A Checklist for the Organization, Operation, and Evaluation of a Company Library*. New York, NY: Special Libraries Association.
This checklist is an excellent guide for establishing any special library. General problems of library management are discussed, followed by an A–Z handbook for setting up the library. Bibliographies are included.

386 **Gorman, M.** 1981. *The Concise AACR2: Being a Rewritten and Simplified Version of the Anglo-American Cataloguing Rules*. 2nd ed. Chicago, IL: American Library Association. 164 p.

387 **Lendvay, O.** 1980. *Primer for Agricultural Libraries*. Wageningen, The Netherlands: Centre for Agricultural Publishing and Documentation. 91 p.
This brief guide covers all aspects of establishing and running a special library. Despite the focus on agriculture, it is equally useful for marine science and aquaculture libraries.

388 **Maxwell, M. F.** 1980. *Handbook for AACR2: Explaining and Illustrating Anglo-American Cataloguing Rules*. Chicago, IL: American Library Association. 463 p.

389 **Morin-Labatut, G.** and **M. Sly.** 1982. *Manual for the Preparation of Records in Developing Information Systems*. Ottawa: International Development Research Centre. (Recommended Methods for Development Information Systems. Vol. 1).
This manual gives detailed explanations and examples for cataloguing and recording information, particularly for automated systems. Volumes 2 and 3 are in preparation.

390 **Oak, L.** 1980. *Special Libraries Bibliographies*. Ottawa: National Library of Canada. Library Documentation Centre. 12 p.
This free bibliography lists key references on establishing and administering a library, including budgeting, cataloguing and library planning.

391 **Olle, J. G.** 1984. *A Guide to Sources of Information in Libraries*. Brookfield, VT, and Aldershot, Hampshire, England: Gower Publishing Co. 178 p.

392 **Ozaki, H.** and **N. Forestell**, comps. 1985. *Special Libraries: Selected References*. Part 1, *Aids for Setting Up a Special Library*. Ottawa: National Library. Library Documentation Centre. 7 p.
Available free of charge from the National Library of Canada.

393 **Petru, W. C.** and **M. W. West**, eds. 1967. *The Library: An Introduction for Library Assistants*. New York, NY: Special Libraries Association.

394 **Rowley, J. A.** 1984. *Organizing Knowledge: An Introduction to Information Retrieval*. Brookfield, VT, and Aldershot, Hampshire, England: Gower Publishing Co.

395 **Wynar, B. S.** 1980. *Introduction to Cataloguing and Classification*. Littleton, CO: Libraries Unlimited, Inc. 657 p.
A simple 'how to' guide to the cataloguing, filing, and classification systems used in libraries.

396 **Zabala, P. T.** 1983. *Guide to the Organization and Management of a Small Fishery Library*. Manila: FAO. South China Sea Fisheries Development and Coordinating Programme. (South China Sea Manuals, No. 3). 86 p.
This manual is an extremely useful guide to the establishing and maintaining of a fisheries and aquaculture library. All areas of library procedures and services are discussed, and examples of classification schemes, form letters, and standard procedures are plentiful.

8 COMPUTERIZED INFORMATION SEARCHES

8.1 Bibliographic computer databases

The use of computers to create bibliographic databases has been a major development in information organization and retrieval. Bibliographic computer databases are computer-searchable indexes to information and are often the equivalent of printed abstract journals. These databases allow researchers efficient access to a huge quantity of information. Database searches, also referred to as computer literature searches or on-line searches, can be useful to researchers in a number of ways. They can generate fairly comprehensive subject bibliographies, or can retrieve a few key references on a topic, all in a matter of minutes. Computer literature searches can be used to verify or supplement bibliographic information for interlibrary loans, cataloguing, or document retrieval. Searches can also quickly provide the researcher with an idea of the amount of information on a subject. A computer literature search can compile a list of references by a particular scientist, or, as with *Science Citation Index* (SCISEARCH), a list of other scientists working in an area of interest. Reference questions can often be answered via a computer literature search, particularly from databases that include textual abstracts or statistical tables. Database searches can also be run regularly to keep abreast of new information. Finally, many databases offer on-line document ordering services; a searcher can conduct a literature search, peruse the results, and order relevant documents without leaving his or her office.

Most databases cover specific subject areas, such as medicine or engineering, but the interdisciplinary nature of modern scientific and technical research often makes it necessary to consider searching databases in parallel or related fields. For example, technicians in the aquaculture industry can find a great deal of useful information on fish processing in the food science and fisheries and agriculture databases, and information on markets, new products, or the activities of their corporate competitors in the business databases. The databases in related fields often provide coverage of sources not found in those covering the subject itself.

Most academic, government, and technical libraries, as well as many of the larger public and corporate libraries in developed countries, have access to the systems of at least one of the major database suppliers. Database suppliers, including Lockheed's DIALOG, the European Space Agency's IRS, EVCONET (Europe), CAN/OLE (Canada), DIMDI (Germany), and IFREMER (France), act as 'hosts', allowing their customers to select from any of the databases they make available.

Libraries or information centres that have access to commercially available databases usually run database searches for their patrons. This service is often offered free of charge or at a subsidized rate in academic, institutional, and some public libraries. Typically the library patron will submit a request for a computer search either by describing the subject of interest in writing, or by discussing the subject with the reference librarian (or staff member trained in database searching). The searcher will select one or a number of databases covering literature on this topic and will formulate a 'search strategy' for each database. A search strategy involves a combination of keywords (subject descriptors, index terms, geographic terms, authors' names, etc.), the logical operators used to string these together (i.e. 'and', 'or', or 'not'), and the command language used by the system supplier. Databases generally require slightly different search strategies depending on the indexing techniques of the database publisher, and the computer command language used by the database supplier.

Once a search strategy has been formulated, the searcher can access the appropriate database, type in the search commands and keywords, and obtain the results of the search. These may be printed on-line, but if a large number of references are retrieved it is cheaper to request the results to be printed off-line. In the latter case, the references retrieved will be printed out at the database supplier's computer facilities and mailed to the searcher. This is a means of cutting on-line computer time costs but it does entail a delay in obtaining search results.

Researchers without access to a local library offering computer literature search services can often request searches by mail from their national science library, or from regional or international development agencies such as the Food and Agriculture Organization, in Rome. (Part III discusses the various services offered by the major marine science institutions.) Another, though more expensive, means of obtaining computer database searches is to use a commercial database search service. These are private companies, sometimes affiliated with a database supplier, which offer computer literature searches, and often document retrieval services as well, for a fee. The charge for their services is based on the database and telecommunications charges incurred during the search, plus an hourly fee or percentage. Document retrieval charges are usually charged per item, which includes photocopying charges and staff time. (Institutions offering this service are discussed in Part III.)

Researchers can also choose to search computer databases themselves. To do so requires an initial investment of both time and money. To begin searching, one needs a computer terminal with communications capabilities (for example, a modem) to connect the computer via the telephone lines to telecommunications networks, and a printer for obtaining on-line search results. Microcomputers are becoming increasingly popular for on-line searching, partly because they are cheap to purchase and therefore widely available. Another advantage of using a microcomputer for searching is that search results can be 'dumped' onto a floppy disk and printed out at the searcher's convenience, which saves on-line costs and provides what in effect is a personal mini-database of references. Most microcomputers require a communications card and program to allow interface between the computer, the telecommunications network, and the database supplier. Database system suppliers and retail computer outlets will generally assist their clients in determining the type of equipment needed to begin on-line searching. This can be a relatively expensive proposition for most researchers, but as the prices of personal computers (microcomputers) continue to drop, and the variety of uses for microcomputers expands, it should be an increasingly viable one. Cooperative arrangements with colleagues or organizations with computer facilities can also provide a means of accessing databases. Renting or leasing equipment from a database supplier or computer outlet might be a practical alternative, particularly for researchers or institutions in the process of determining staff demands and requirements for on-line searches.

The next step is to arrange a contract with a database supplier. These contracts basically state that the user agrees to pay the costs incurred when he searches the databases, and that he will respect the copyright, or any other usage restrictions, stipulated by the database publishers. Some suppliers charge a fee to initiate service, or have an annual subscription charge. If charges are imposed check first to see if the databases of interest are available through other suppliers. Many of the major databases, such as BIOSIS or Chemical Abstracts, are available through a number of suppliers' systems; compare the pricing structure for the primary databases in your area before deciding on one system. It is often more efficient to search some databases on one system and others on a different system simply because of the differences in contract and database charges.

Once the contract has been arranged, the searcher will be assigned a unique password which allows him access to the system's databases. Database suppliers regularly schedule inexpensive training courses in major cities worldwide. These courses explain search strategies and techniques, command language, and the special features of the databases they provide. The system manual and the guide manuals for key databases should also be purchased. These allow the searcher to use the databases more efficiently and effectively, and their cost will quickly be recovered in saved 'on-line' time. Would-be searchers who cannot attend a training session can still run computer searches either simply through self-instruction using the system's manual, or by purchasing a 'gateway' software program for their microcomputer. Gateway programs, for example 'In-Search' (produced by the Menlo Corporation) for the DIALOG system, provide on-line user assistance for database searching. 'Training' databases are also available on some systems. These are mini-versions of the major databases and are priced at an inexpensive hourly rate to allow searchers to practise online searching. (For example, the ONTAP databases in the DIALOG system are instructional databases). Suppliers also offer free user assistance to their clients, including advice on choosing the appropriate database, and on formulating a search strategy.

When to Use Computer Literature Searches

Databases are proliferating at a tremendous rate and almost all areas of science are covered by at least one major database. Computer literature searches are most effective for research requiring current, published information. Though databases do collectively offer a considerable range of coverage, in both time and content, most do not presently extend coverage beyond the last twenty years. Moreover, coverage of non-published research, particularly of consultants', government, and corporate studies, is generally not as comprehensive as that of journals and published monographs. Despite these considerations on-line literature searches are an increasingly indispensable means of finding scientific information.

Database searching offers the researcher a number of advantages over traditional literature searches of printed indexes and catalogues. One main advantage is timeliness. The lag time between when a document is indexed and when the document is accessible through the index is much shorter in a computer database than in a printed (published) index. A second advantage is that the computer can search for a combination of index terms in seconds, saving the researcher hours of ploughing through printed indexes, checking under numerous subject headings. Most databases provide greater access to information than do traditional printed indexes as the computer can search for words or phrases in the title and abstract, as well as the assigned keywords, descriptors, and authors' names. Moreover, the computer print-out of the search results gives the researcher an instant recorded bibliography, eliminating the need for hand copying or photocopying of citations from printed indexes. An on-line search that takes less than an hour to prepare and run could take days to compile using printed indexes.

It should be noted that although many databases do have a printed equivalent an increasing number do not. In many cases the *only* access to a large quantity of information is via an on-line search. This state of affairs will likely continue as printed indexes become more costly to publish. Computer literature searches can also be extremely cost-effective for smaller libraries that cannot afford to purchase expensive printed indexes. By using the equivalent on-line database, libraries can still offer their users access to the information, and pay only for what they actually use. This consideration can be particularly important for retrieving information on subjects outside the main focus of the library collection.

On-line searching can also assist researchers in keeping up to date in their fields. Most systems have the capability to 'save' or 'store' search strategies, which can then be run again at regular intervals to obtain new references added to the database since the initial search. A number of databases have a facility for automatically running the search strategy as the database is updated. This is often referred to as Selective Dissemination of Information or SDI. To establish an SDI profile a searcher merely instructs the system to save a particular search strategy as an SDI. As the particular database is updated the search will be run again on the 'update' segment and a printout of new references will be mailed to the searcher.

The wealth of scientific literature and the variety of originating sources make computer literature searches an invaluable tool for locating marine science information. They provide an excellent base of literature that can be supplemented by the use of other literature search methods, particularly to obtain pre-database and unpublished documents.

8.2 Aquaculture databases

Aquaculture literature is extensively covered by two major databases, Aquatic Sciences and Fisheries Abstracts (ASFA) (and its sub-database, Aquaculture) and Oceanic Abstracts (OA). Both include all fields of aquaculture, the major difference being ASFA's additional coverage of freshwater sciences. Other databases to be considered are some of the major agricultural databases, such as AGRICOLA or Commonwealth Agricultural Bureaux (CAB) Abstracts (both of which include some information on the culture of aquatic plants and organisms), and BIOSIS. Though there may be some duplication of references, databases in related fields usually cover a different set of journal and monograph sources. Thus they can enable the searcher to retrieve literature published in sources not generally considered by major marine science and fisheries databases. Researchers in the fishing and fish processing industries will find the agricultural (CAB and AGRICOLA) and the food processing databases (for example, Food Science and Technology Abstracts (FSTA) and FOODS ADLIBRA) to be excellent sources of information in their field. Chemical Abstracts can also be a useful source of information on the chemical processes involved in fish processing, including food contamination and mycotoxins. Aquaculture engineers might find searches of Engineering Index, in addition to searches of ASFA and Oceanic Abstracts, valuable.

Multidisciplinary databases are also good sources of aquaculture information. National Technical Information Service (NTIS) is a database that covers government (mainly, but not exclusively United States) sponsored research in all areas of science. It has useful information on aquaculture, marine science research, marine transportation, marine pollution, invertebrate biology, and fish marketing and processing. SCISEARCH (*Science Citation Index*) is another multidisciplinary database that can be particularly useful for finding references by a scientist, or to determine who has cited works by a particular author—in short, to find who is conducting research in a specialized area.

Following a listing of sources of general information on computer databases is a selection of the main databases covering aquaculture literature. Researchers should write to the major database suppliers to obtain free catalogues describing their databases and the subject areas covered. As mentioned above, searchers should always consider database searches of related subject fields, to supplement searches of the primary aquaculture and fisheries databases.

397 Blanchard, J. R. and **L. Farrell**, eds. 1981. *Guide to Sources for Agricultural and Biological Research*. Berkeley, CA: University of California Press. 735 p.
This text includes an excellent summary on the development of bibliographic databases and an explanation of how to use databases for bibliographic searches.

398 Byerly, G. 1983. *Online Searching: A Dictionary and Bibliographic Guide*. Littleton, CO: Libraries Unlimited, Inc.
Online Searching is a useful guide combining a dictionary of database and computer terminology with an extensive, abstracted bibliography of key articles on on-line searching and the use of bibliographic databases.

399 Casbon, S. 1983 'Online searching with a microcomputer—getting started'. *ONLINE* 7 (November): 42–46.
This article explains the advantages and disadvantages of using a microcomputer for on-line searching. A helpful bibliography of further readings is also included.

400 Chen, C. and **S. Schweizer.** 1981. *Online Bibliographic Searching: A Learning Manual*. New York, NY: Neal-Schuman Publishers; London: Mansell. (Applications in Information Management and Technology series). 244 p.
A straightforward guide to using on-line databases, including sample searches, a directory of databases and suppliers, and an extensive bibliography.

401 *Database*. 1978–. Quarterly. En. Weston, CT: Online, Inc.
A quarterly journal devoted to practical articles on the use of databases.

402 Deutsch, M. 1984. *Online Databases*. Burnaby, British Columbia: Simon Fraser University Library. 4 vols.
An excellent and inexpensive guide to bibliographic and some statistical databases. (Coverage includes only North American suppliers.)

403 *Directory of Online Databases*. 1984–. Santa Monica, CA: Cuadra Associates.
Updated twice yearly. This is a useful directory of bibliographic and nonbibliographic databases.

404 *Directory of Online Information Resources: A Guide to Commercially Available Data Bases*. 1983. Kensington, MD: Capital Systems Group, Inc.
A regularly updated directory of bibliographic databases. Entries include information on coverage, file size, costs, and an abstract describing contents. Publishers and vendors, with addresses, are also included.

405 **Ewbank, W. B.** 1982. 'Comparison guide to selection of databases and database services'. *Drexel Library Quarterly* 18 (3/4): 189–204.
This article is a practical explanation of how to evaluate databases.

406 *Going Online 1984*. 1984. London: Online Information Centre, Aslib.
An excellent introductory guide to online computer database searching. Topics covered include equipment, copyright, suppliers addresses, and online user groups (particularly European groups).

407 **Hansen, C.** 1984. *The Microcomputer User's Guide to Information Online*. Hasbrouck Heights, NJ: Hayden Book Co.
This text is a simple guide to online searching the databases available and the hardware and software needed for computer database searches.

408 'International comparative price guide to databases online'. 1984. *Online Review* 8 (1): 105–112.
An annual review of major database suppliers' charges.

409 **Levy, L. R.** 1984. 'Gateway software: is it for you?'. *Online* 8 (6): 67–79.
An explanation of the capabilities of gateway software and an evaluation of the major programs available.

410 **Lilley, G. P.**, ed. 1981. *Information Sources in Agriculture and Food Science*. London: Butterworth. (Butterworths Guides to Information Sources). 618 p.
The basics of on-line searching techniques are discussed. The discussion of agricultural and food science databases is also useful for researchers involved in aquaculture and fish processing.

411 **Newlin, B.** 1985. *Answers Online: Your Guide to Informational Databases*. Berkeley, CA: Osborne McGraw-Hill.
A useful introduction to database searching, including explanations of the computer hardware and software required, hints for searching, sample searches, and special services available online, such as electronic mail.

412 *ONLINE*. 1976–. Bimonthly. En. Weston, CT: Online, Inc.
Covers computerized information science, in particular online searching and library automation. It is an excellent source of information concerning new databases, techniques for searching, and current publications.

413 **Palmer, R. C.** 1983. *Online Reference and Information Retrieval*. Littleton. CO: Libraries Unlimited, Inc. (Library Science Text series).

This book offers a simple guide to database structure, suppliers' systems, computer equipment needed, and user aids. The many sample searches described make the explanations especially easy to understand.

414 *R & D Database Handbook: A Worldwide Guide to Key Scientific and Technical Databases*. 1984. Fort Lee, NJ: Technical Insights, Inc.
This directory is an international listing of over 500 scientific and technical databases. Included are suppliers' addresses, descriptions of the types of information in each database, and a subject index.

415 **Starr, S. S.** 1982. 'Databases in the Marine Sciences.' *Online Review* 6 (2): 109–125.
This article is a useful examination of the contents and indexing techniques of Oceanic Abstracts, Aquatic Sciences and Fisheries Abstracts, BIOSIS and GEOREF.

416 **Williams, M. E., L. Lannom** and **C. G. Robins**. 1982. *Computer-Readable Databases: A Directory and Data Sourcebook*. Washington, DC: American Society for Information Science. 1,472 p.
A comprehensive listing of over 700 databases. International coverage is good and each citation includes an abstract describing content and coverage.

Major Bibliographic Databases Covering Aquaculture Literature

417 **AGRICOLA.** Science and Education Administration, Technical Information Systems (SEA/TIS), U.S. Department of Agriculture, National Agricultural Library Building, Beltsville, MD 20705, U.S.A. Printed equivalents: *Bibliography of Agriculture*, *The National Agricultural Library Catalog* and the *Catalog of the Food and Nutrition Information and Education Resources Center*. Time span: 1970–present.
AGRICOLA is the library catalogue and index issued by the U.S. National Agricultural Library. It provides comprehensive international coverage of journal and monograph literature on agriculture and related subjects, including fisheries and (up to 1985) aquaculture. Since 1985 aquaculture information has been input into the ASFIS document delivery service, available worldwide for aquaculture-related literature.

418 **AGRIS INTERNATIONAL.** FAO, AGRIS Processing Unit, IEA, P.O. Box 100, A-1400 Vienna, Austria. Printed equivalent: *AGRINDEX*. Time span: 1975–present.
AGRIS indexes references contributed by over 100 national and multinational centres working in the fields of agriculture, fisheries, and aquaculture. It is particularly useful for non-U.S. materials, and for research originating in developing countries.

419 **AQUACULTURE.** U.S. National Oceanic and Atmospheric Administration (NOAA), Chief, User Services Branch, NOAA/EDIS/ESIC/-LISD, 6009 Executive Boulevard, Rockville, MD 20852, U.S.A. No printed equivalent. Time span: 1970–January 1984.

AQUACULTURE provides coverage of information on the culture of marine, brackish-water and freshwater organisms. Monographs, periodicals, and conference proceedings are included. Subjects covered range from economic and legal aspects of aquaculture to the physical, biological, and environmental requirements for the culture of aquatic organisms. This database was closed in January 1984 but is available for retrospective searches.

420 **AQUATIC SCIENCES AND FISHERIES ABSTRACTS** (ASFA). Fishery Information, Data and Statistics Service, Food and Agriculture Organization of the United Nations, Via delle Terme di Caracalla, I-00100 Rome, Italy. Printed equivalents: (1) *Aquatic Sciences and Fisheries Abstracts*. Part 1: *Biological Sciences and Living Resources*; Part 2: *Ocean Technology, Policy, and Non-Living Resources*; (2) *Aquaculture Abstracts*. Time span: 1978–present.

ASFA is the major international database for freshwater and marine sciences. It is produced by an international body and coordinated by FAO list partners. It includes information on aquatic biology, oceanography, fisheries, aquaculture, water pollution, marine technology, marine geosciences, maritime law, and socioeconomic research on aquatic resources. Over 5,000 primary journals and other source documents are regularly monitored for inclusion in ASFA. Since 1986 the ASFA database has been available on compact disk.

421 **BIOSIS PREVIEWS.** BioSciences Information Service, 2100 Arch Street, Philadelphia, PA 19103–1399, U.S.A. Printed equivalents: *Biological Abstracts*, *Biological Abstracts/Reports, Reviews, Meetings*, and *Bioresearch Index*. Time span: 1969–present.

BIOSIS provides the most comprehensive worldwide coverage of research in the life sciences, including aquaculture. It is probably the most widely used database for aquatic biology in particular. Over 9,000 journals and monographs, as well as conference proceedings, reviews, and government and institutional reports are regularly monitored to provide extensive coverage of all aspects of bioscience.

422 **COMMONWEALTH AGRICULTURAL BUREAUX ABSTRACTS (CAB ABSTRACTS).** Commonwealth Agricultural Bureaux (CAB), Farnham House, Farnham Royal, Slough SL2 3BN, England. Printed equivalents: twenty-six CAB journals. Time span: 1972–present.

CAB ABSTRACTS contains all the records published in the twenty-six major abstract journals of the Commonwealth Agricultural Bureaux. Over 8,500 journals in thirty-seven languages are monitored, as well as monographs, reports, and other publications. Major references are abstracted, other publications are referenced with a bibliographic citation. CAB's coverage of literature on developing nations is particularly useful.

423 **DISSERTATION ABSTRACTS ONLINE.** University Microfilms International, Dissertation Publishing, 300 North Zeeb Road, Ann Arbor, MI 48106, U.S.A. Printed equivalent: *Dissertation Abstracts International*. Time span: 1861–present.

DISSERTATION ABSTRACTS provides author, subject, and title access to all U.S. doctoral dissertations since 1861. Master's theses have been selectively included since 1962, and an increasing number of foreign dissertations are also covered.

424 FOOD SCIENCE AND TECHNOLOGY ABSTRACTS (FSTA). International Food Information Service (IFIS), Arabella Center, Lyoner Strasse 44–48, D-6000 Frankfurt am Main 71, Federal Republic of Germany. Printed equivalent: *Food Science and Technology Abstracts*. Time span: 1969–present.

FSTA covers research and development in the field of food science. Related disciplines, for example agriculture, biochemistry, and physics, are also covered. Over 1,200 journals are monitored, as well as patents from twenty countries, monographs, and other reports.

425 FOODS ADLIBRA. General Mills Foods Adlibra Publications, 9000 Plymouth Avenue North, Minneapolis, MN 55427, U.S.A. Printed equivalent: *Foods Adlibra*. Time span: 1974–present.

FOODS ADLIBRA covers the latest developments in food processing and technology. It includes information on retailers, brokers, processors, suppliers, and importers, and is a useful source of information on government guidelines and regulations on the processing and packaging of foods. Marketing and management news, statistics, and information on food production economics are also covered.

426 NATIONAL TECHNICAL INFORMATION SERVICE (NTIS). National Technical Information Service (NTIS), U.S. Department of Commerce, 5285 Port Royal Road, Springfield, VA 22161, U.S.A. Printed equivalents: several publications, including *Government Reports, Announcements and Index*, and twenty-six abstracts newsletters. Time span: 1964–present.

The NTIS database provides access to U.S. government-sponsored research. This includes coverage of reports issued by federal agencies, or by their contractors. Reports on aquaculture, climatology, meteorology, food processing, marine biology, and marine geosciences are all covered. Non-U.S. government research reports that are available for purchase from NTIS are also included.

427 OCEANIC ABSTRACTS (OA). Cambridge Scientific Abstracts, 5161 River Road, Bethesda, MD 20016, U.S.A. Printed equivalent: *Oceanic Abstracts*. Time span: 1964–present.

OA is an international marine science database recording references from journals, books, technical reports, conferences, and government and trade publications. The subject coverage includes oceanography, marine biology, pollution, maritime transportation, geosciences, and maritime law. OA has a more selective policy for data inclusion than does ASFA or BIOSIS. This policy assists the user in retrieving major articles on a topic, but will likely result in a less comprehensive search.

428 SCIENCE CITATION INDEX (SCISEARCH). Institute for Scientific Information, University City Science Center, 3501 Market Street, Philadelphia, PA 19104, U.S.A. Printed equivalent: *Science Citation Index*. Time span: 1974–present.

A multidisciplinary citation index covering all fields of scientific and technical information worldwide.

429 **VETOMER.** Laboratoire de Pathologie des Animaux Aquatiques, Ministère de l'Agriculture, INFREMER—Centre de Brest, B.P. 337, F-29273 Brest Cedex, France. No printed equivalent. Time span: 1976–present.
VETOMER specializes in references on the pathology of marine and freshwater fish, including infections, environmental and nutritional diseases, and disease control.

430 **ZOOLOGICAL RECORD (ZR).** BioSciences Information Service, 2100 Arch Street, Philadelphia, PA 19103-1399, U.S.A. Printed equivalent: *Zoological Record.* Time span: 1978–present.
ZR is an international database of zoological literature with twenty-seven sections covering various animal groups: protozoa, nematoda, pisces, reptilia, aves, mammalia, etc.

Database Suppliers

431 **DIALOG Information Services, Inc.** 3460 Hillview Avenue, Palo Alto, CA 94304, U.S.A.
Tel: (415) 858-2700 Telex: 334499
Start-up fee: No. Monthly minimum charge: No. Discount options: Yes. SDI (Selective Dissemination of Information) service: Yes.

432 **ESA/IRS.** European Space Agency, Via Galileo, C.P. 64, I-00044 Frascati (Roma), Italy.
Tel: + 39 6 94011
Start-up fee: No. Monthly minimum charge: No.

433 **System Development Corporation (SDC).** Information Services (Orbit Search Service Databases), 2500 Colorado Avenue, Santa Monica, CA 90406, U.S.A.
Tel: (213) 820-4111; 800/421-7229 (toll-free in continental U.S.A.); 800/352-6689 (toll-free in California); (213) 453-6194 (collect calls accepted from Canada) Telex: 65-2358 TWX: 910/343-6643
Start-up fee: No. Monthly minimum charge: Yes (with some exceptions). Discount options: Yes. SDI (Selective Dissemination of Information) service: Yes.

434 **CAN/OLE CAN/SDI.** Client Services, CISTI, National Research Council of Canada, Ottawa, Ontario K1A 0S2, Canada.
Tel: (613) 993-1210 Telex: 053-3115 Envoy 100: CISTI.CLIENT.SERV
Start-up fee: No. Monthly minimum charge: No. Discount options: No. SDI (Selective Dissemination of Information) service: Yes.

435 DIMDI. Weishausstrasse 27, Postfach 420580, D-5000 Cologne 41, Federal Republic of Germany.
Tel: + 49 221 47241
Start-up fee: No. Minimum charge: Yes (on a quarterly basis).

436 Bibliographic Retrieval Service (BRS), Inc., 1200 Route 7, Latham, New York 12110, U.S.A.
Tel: (518) 783-1161; 800/833-4707 (toll-free in U.S.A.); 800/553-5566 (toll-free in New York State); (518) 783-7251 (collect from Canada) TWX: 7104444956
Start-up fee: No. Monthly minimum charge: No. Discount options: Yes. SDI (Selective Dissemination of Information) service: Yes.

437 Institut Français de Recherche pour l'Exploitation de la Mer (IFREMER), B.P. 337, F-29273 Brest Cedex, France.

Non-Bibliographic Databases

438 AQUADOC. Bureau National des Données Océaniques (BNDO), IFREMER (address as in [437]). No printed equivalent. Time span: Current year.
AQUADOC is a directory of French companies and organizations working in the field of aquaculture.

439 GLOBEFISH. FAO. No printed equivalent.
GLOBEFISH is a marketing database particularly designed to assist developing countries in determining short- and medium-range markets for fishery products. The INFOFISH and INFOPESCA offices (*see* [473C]) provide services to interested users who do not have access to the network through their own computer facilities.

9 CURRENT AWARENESS

Most aquaculturalists make some attempt to keep abreast of new developments in their areas of specialization. However, the volume of published scientific literature can be overwhelming, and an individual can spend a great deal of time scanning through the numerous potential sources of information. A centralized 'current awareness' service operated by the library or information centre is the most cost-effective means of ensuring that staff are kept aware of current research. Although the range of services will vary according to the role of the library in the parent organization, most libraries offer some form of current awareness service to their patrons.

Current-awareness or 'selective dissemination of information' services entail monitoring a variety of sources of new publications and circulating the relevant information to the researchers working in the relevant subject area. This service is of mutual benefit to library users and library staff. Library users are informed of new literature or work of interest to them, and the library promotes itself as an active, current centre for research information.

The following categories of information are usually good sources of current information:

- current journals;
- newsletters and irregular serials;
- bibliographic computer databases;
- annuals or yearbooks;
- conference and workshop proceedings;
- publications and accessions lists from other associations, institutions and libraries;
- publishers' catalogues; and
- abstracting and indexing services.

In fact, many of these types of source developed specifically to meet the need for knowledge of current scientific research.

In the following bibliography sources discussed elsewhere in this handbook have not been included. Readers should refer to the appropriate sections for information on journals, computer database searches, institutions, and abstracts and indexes.

9.1 Newsletters

440 *Aquarius*. 1977–. 3 per year. Es, En. Lima: Comité de Acción de Productos del Mar y de Agua Dulce del Sistema Económico Latinoamericano.
Aquarius is the newsletter of OLDEPESCA. It provides summaries of workshops and meetings related to the development of Latin American fisheries, listings of new publications, and news of scientific research in the region.

441 *Artemia Newsletter*. Quarterly. En. Ghent: Artemia Reference Centre, State University of Ghent.
This newsletter provides information on courses and meetings, information, abstracts of selected papers on aquaculture, and bibliographies.

442 *ASFA Newsletter*. 1974–. Irregular. En. Plymouth: Marine Biological Association, for FAO.
Aquatic Sciences and Fisheries Abstracts Newsletter covers items related to ASFA activities and areas of interest, including personnel appointments, training, and institutional news.

443 *Bay of Bengal News*. En. Madras, India: Bay of Bengal Programme.
This is the publication of the SIDA/FAO/UNDP Bay of Bengal Programme for Fisheries Development. News items on the fisheries of the countries bordering the Bay of Bengal are described.

444 *CECAF Newsletter*. Irregular. En. Dakar, Senegal: FAO/UNDP Project for the Development of Fisheries in the Eastern Central Atlantic.

The *CECAF Newsletter* discusses current research on the fishery resources of the Eastern Central Atlantic, and, in particular, the status of FAO/UNDP Projects in the area.

445 *A Current Awareness Bibliography for IDRC-Supported Fisheries Projects*. 1977–82. Quarterly. En, Fr, Es. Ottawa: International Development Research Centre.
This quarterly bibliography listed new references related to the fisheries work carried out by IDRC staff in developing countries. References were listed alphabetically by author with additional taxonomic and subject indexes.

446 *Eurofish Report*. Biweekly. En. Tunbridge Wells, Kent, England: Agra Europe (London).
This newsletter reviews the European fishing industry and discusses catch statistics, processing trends, consumer markets, government legislation, and new research developments.

447 *European Mariculture Society Newsletter*. Quarterly. En. Bredene, Belgium: European Mariculture Society.
News items of interest to mariculturists worldwide, including workshops, publications, advertisements for equipment and services, and training courses, are given in this useful publication.

448 *Fishbyte: Newsletter of the Network of Tropical Fisheries Scientists*. 1984–. Makati, Metro Manila, The Philippines: International Center for Living Aquatic Resources Management.

449 *ICLARM Newsletter*. 1977–. Quarterly. En. Makati, Metro Manila, The Philippines: International Center for Living Aquatic Resources Management.
ICLARM Newsletter is an excellent source of information on current fisheries and aquaculture research in tropical developing countries. Scientific articles, directories of agencies, news items, book reviews, and subject bibliographies of new literature are regularly included.

450 *IMS [International Marine Science] Newsletter*. 1973–. Quarterly. En, Fr, Es. Paris: *UNESCO*, Intergovernmental Oceanographic Commission.
The *IMS Newsletter* is published by IOC/UNESCO in collaboration with the UN, FAO, IMO, and WMO. It includes news items on the international marine science activities of these organizations.

451 *INFOFISH Trade News*. Biweekly. En. Kuala Lumpur: INFOFISH.
This biweekly newsletter summarizes market information, including prices, for fisheries products.

452 *INFOPESCA*. Biweekly. Es. Panama City: INFOPESCA.
This bulletin covers regional and international marketing information for fish products from Latin America.

453 *International Association of Marine Science Libraries and Information Centres (IAMSLIC) Newsletter*. Irregular. En. Fort Pierce, FL: IAMSLIC (c/o Kristen L. Metzger, Harbor Branch Foundation).
Available to members of IAMSLIC, this newsletter is an excellent source of information on marine science information centers, databases, research, and publications.

454 *NACA Newsletter*. 2 per year. En. Iloilo City, The Philippines: Network of Aquaculture Centres in Asia.
This newsletter describes the current projects, programmes and activities of the Network of Aquaculture Centres in Asia (NACA).

455 *SEAFDEC Newsletter*. 1977–. Quarterly. En. Bangkok: Southeast Asian Fisheries Development Center.
Upcoming meetings, new publications, feature articles, notices of appointments, and training courses are regularly covered in this fisheries newsletter for the Southeast Asia region.

456 *SEANET Market Newsletter*. 1985–. Biweekly. En. Camden, ME: Journal Publications.
Produced by the publishers of *Seafood Business Report* and *National Fisherman*, *SEANET* combines a biweekly (paper copy) newsletter with a computerized database, providing up-to-date market statistics and prices for the fishing industry.

457 *South Pacific Commission. Fisheries Newsletter*. Quarterly. En. Noumea, New Caledonia: South Pacific Commission.
This free newsletter describes the activities of the Commission and other items related to fisheries in the South Pacific, including research results, topical articles, conference announcements and staff changes.

458 *Tropical Fish Processing Newsletter*. 1980–82. Irregular. En, Es. Bogotá, Colombia: International Development Research Centre.
This newsletter provided a medium for fish processing technologists to communicate the results and status of their work to their colleagues.

459 *World Aquaculture News*. 1988. Monthly. En. Arlington, Tx: IMS.
This newsletters highlights aquaculture news items and includes information on courses, meetings, publications.

460 **World Aquaculture Society**. Quarterly. En. 16E Fraternity Lane, Louisiana State University, Baton Rouge, LA 70803, U.S.A.
The purpose of this newsletter is to develop communications between all members of the World Aquaculture Society. The newsletter prints items relating to the Society itself, in addition to research updates, recent publications, book reviews, announcement of cruises, meetings, courses, and other information of interest to the members of the Society.

9.2 Annuals, yearbooks

(See also Part III, Organizations, for annual reports of research organizations, and Section 6.9, Statistical Sources).

461 **Barnes, M.,** ed. 1980–. *Oceanography and Marine Biology: An Annual Review.* Oxford: Pergamon Press. (Oceanography and Marine Biology series).

462 **Borgese, E. M.** and **Ginsburg, N.,** eds. 1978–. *Ocean Yearbook.* Chicago, IL: University of Chicago Press.
This yearbook covers current developments in oceanography, including living and nonliving resources, marine transportation, environment, and coastal management. The appendices are particularly useful for summaries of the activities of international oceanographic organizations, and directories of key institutions and personnel.

9.3 Conference and workshop proceedings

The following citations do not form a complete list of aquaculture proceedings, but rather will provide the reader with an idea of the scope of published conference materials available. Information on *upcoming* conferences can be found in the journal and newsletter sources, and in publications of research organizations.

463 *Aquaculture Economics Research in Asia: Proceedings of a Workshop Held in Singapore, 2–5 June, 1981.* 1982. Ottawa: International Development Research Centre. (IDRC, 193). This report covers the proceedings of a workshop on the economic and social importance of aquaculture to Southeast Asia.

464 *Fish Quarantine and Fish Diseases in Southeast Asia: Report of a Workshop Held in Jakarta, Indonesia, 7–10 December, 1982.* 1983. Ottawa: International Development Research Centre. (IDRC, 210).
These proceedings cover a workshop co-sponsored by IDRC and the UNDP/FAO South China Sea Fisheries Development and Coordinating Program. The topics discussed include fish disease research, legislation, quarantine, and certification procedures in Southeast Asia.

465 *Fisheries Information Science in Southeast Asia: Proceedings of a Seminar Held in Bangkok, Thailand 16–20 August, 1982.* 1982. Bangkok: Southeast Asian Fisheries Development Center. (SEAFDEC FIS/82 INF 1. Rev. 1). 2 vols.

466 **Juario, J. V., R. P. Ferraris**, and **L. V. Benitez**, eds. 1984. *Advances in Milkfish Biology and Culture.* Metro Manila, The Philippines: Island Publishing House and IDRC/SEAFDEC. 243 p.
Proceedings of the Second International Milkfish Aquaculture Conference, 4–8 October 1983, Iloilo City, the Philippines.

467 **McNeil, Wm. J.** 1968. *Marine Aquiculture.* Corvallis, OR: Oregon State University Marine Science Center. 172 p.

Selected papers from the Conference on Marine Aquiculture, Oregon State University Marine Science Center, Newport, Oregon, 23–24 May 1968.

468 Pillay, T. V. 1977. *Planning of Aquaculture Development: An Introductory Guide.* Rome: FAO. 71 p.
A selection of papers from the FAO/NORAD Round Table on the 'Strategy for Development of Aquaculture as an Industry', 1974, are included in these proceedings. The Round Table discussed the organization of the aquaculture industry, projections for future growth and development, and the relationship between the industry and the external economic, political, and natural environment.

469 Pillay, T. V. and **W. A. Dill**, eds. 1976. *Advances in Aquaculture.* Farnham, Surrey, England: UK: Fishing News Books. 653 p.
This collection of 117 papers from the FAO Aquaculture Conference held in Kyoto, Japan, includes details of research on fresh and seawater aquaculture, genetics, the culture of crustaceans, molluscs, algae, and seaweeds, and the nutritional requirements of a variety of fish species.

470 Pullin, R. S. and **R. H. Lowe-McConnell**, eds. 1982. *The Biology and Culture of Tilapias.* Manila: ICLARM. 432 p.
Proceedings of the International Conference on the Biology and Culture of Tilapias, 2–5 September 1980, held at the Study and Conference Center of the Rockefeller Foundation, Bellagio, Italy, and sponsored by ICLARM.

471 Pullin, R. S. and **Z. H. Shehadeh**, eds. 1980. *Integrated Agriculture-Aquaculture Farming Systems.* Manila: ICLARM.
Proceedings of the ICLARM–SEARCA Conference on Integrated Agriculture-Aquaculture Farming Systems, 6–9 August 1979.

472 Taki, Y., J. H. Primavera, and **J. A. Llobrera**. 1985. *Proceedings of the First International Conference on the Culture of Penaeid Prawns/Shrimp.* Iloilo, The Philippines: SEAFDEC.

PART III

Directory of Selected International, Regional, and National Organizations

Selected International, Regional, and National Organizations

10 INTERNATIONAL AND REGIONAL ORGANIZATIONS

10.1 Introduction

The most valuable information resources can result from the informal or formal development of exchange links for information and publications among information centres and libraries. This section has been designed to provide details concerning organizations and institutions which offer aquaculture information services either worldwide or on a regional basis. Only centres which offer information services to external users and the general public have been included; centres that do not offer information services have been omitted.

The institutions here are the key organizations that should be contacted by all aquaculture information centres requiring information from many geographical areas. In particular, the services of FAO and ICLARM should be tapped for information services on aquaculture. The services and publications of national organizations that provide information services to external users are described in Section 11.

Excellent descriptions of the information centres and services in marine sciences, fisheries, and aquaculture of the international organizations and those found in Asia and the Pacific can be found in *ICLARM Newsletter*, Volume 5, No. 2, April 1982, and Volume 7, No. 2, April 1984; and in *Fisheries Information Science in Southeast Asia*, Proceedings of Seminar Held in Bangkok, Thailand, 16–20 August 1982 (available from SEAFDEC [480]).

The *Directory of Marine Sciences Libraries and Information Centers* (1981) [60] and its *Supplement* (1984), compiled by Carolyn P. Winn of Woods Hole, provides an excellent list and description of marine sciences, including aquaculture libraries and information centres in the United States, Canada, Bermuda, Fiji, Panama, the United Kingdom, Argentina, Australia, India, Kuwait, Monaco, the Philippines, and South Africa. The information centres included in this directory and supplement are members of the International Association of Marine Science Libraries and Information Centres. Each of them has a policy of assisting any library or information centre seeking information on marine sciences.

A format similar to that used in the above-mentioned *Directory of Marine Science Libraries* to describe the services provided by the centres listed has been adopted in this *Keyguide*. An indication of the services offered and the subject scope is given, along with details of any publications produced, the working language(s) used, and a description of the collection. The centres listed in Winn's directory that offer services internationally have also been listed in the present work.

10.2 International organizations

The Food and Agriculture Organization of the United Nations (FAO)

473A International Information System for the Agricultural Sciences and Technology (AGRIS), AGRIS Coordinating Centre, FAO, Via delle Terme di Caracalla, I-00100 Rome, Italy
Tel: (06) 57991 Telex: 610181 FAO Cable: FOODAGRI Rome

AGRIS is an international information system that collects and indexes agricultural literature among its member countries and facilitates information exchange among them. FAO coordinates this activity. The AGRIS Input Unit is in Vienna.

Users Policy: Open to public.

Services: AGRINDEX is available in hard copy and can be searched in the AGRIS database. Each year 100,000 citations are added to the database. SDI services and bibliographies are offered. Regional databases have been established.

Subjects: Subject groups covered in AGRIS include: aquatic sciences, fisheries, aquaculture, marine products, human nutrition, pollution, and economics.

Publications: *AGRINDEX* (monthly, no abstracts), over 120,000 citations per year; *AGRIS: Guide for Users*.

Regional databases have been established. Included are:

(a) Agricultural Information Bank for Asia (AIBA), SEARCA, MCC P.O. Box 720, Makati, Metro Manila, The Philippines.
Contact: Mrs. J. C. Sison

Has eight member states from Asia. Maintains in-house database, AGRIASIA, which is also published in hard copy. A fisheries and aquaculture publication drawn from citations in *AGRIASIA* has been published periodically starting from 1982. Publications are available to any interested users. Offers database searching, retrieval services, SDI services, searches in *AGRIASIA* and

AGRINDEX, document delivery, training in information processing and management.

Publications: *AGRIASIA* (no abstracts), *Fisheries Bibliography* (no abstracts).

(b) Commission of the European Communities (CEC), Brussels, Belgium
Based on input from the EEC member states.

(c) Centro Interamericano de Documentación e Información y Comunicación Agrícola (CIDIA), Costa Rica
All Latin American and Caribbean countries, except Brazil and Cuba, input into CIDIA. It maintains the database AGRINTER, which is also published in hard copy. Offers the same services as AIBA in Asia.

473B Current Agricultural Research Information System (CARIS), Caris Coordinating Centre, FAO, Via delle Terme di Caracalla, I-00100 Rome, Italy
Based on input from national input centres for AGRIS, information is collected on agriculture (including fisheries and aquaculture) institutions, scientists, programmes, and projects worldwide.

473C Fishery Information, Data and Statistics Service, Fisheries Department, FAO, Via delle Terme di Caracalla, I-00100 Rome, Italy
Tel: 579-767-088
Contact: Chief, Information
The fisheries information programmes of FAO include:

a. Aquatic Sciences and Fisheries Information System (ASFIS).
b. Aquaculture Development and Coordination Programme (ADCP).
c. *Marine Science Contents Tables* (*MSCT*).
d. *Freshwater and Aquaculture Contents Tables* (*FACT*).
e. GLOBEFISH
f. Indo-Pacific Fishery Commission (IPFC).
g. Tuna Resources Development and Management in the Indo-Pacific (Reg Tuna).
h. Bay of Bengal Programme (BOBP).

We shall examine each of these in turn.

(a) Aquatic Sciences and Fisheries Information System (ASFIS)

ASFIS is co-sponsored by the Ocean Economics and Technology Board (OETB) of the UN Department of International Economic and Social Affairs, the Intergovernmental Oceanographic Commission (IOC) of UNESCO, and FAO. The major product of ASFIS, *Aquatic Sciences and Fisheries Abstracts* (ASFA), is published by Cambridge Scientific Abstracts, Washington, DC.

Users Policy: ASFA services are available to the public through database suppliers such as DIALOG (see Section 8.2). Many of the subscribers to ASFA are from developing countries.

Services Available to Public: Monthly updates to the information system are published in *Aquatic Sciences and Fisheries Abstracts* (on-line or in hard copy). *ASFIS*

Experts Register is published on an *ad hoc* basis. It includes a bibliographic index of fishery scientists worldwide. The *ASFIS Thesaurus* will be available in the near future. Photocopying (fee), interlibrary loans (FAO documents only), and numerical data services are offered. Fish catch and landing statistics, fishery commodities statistics, fishery vessel statistics, computer database searching (FAO Documentation Database, ASFIS, AGRIS, FISHDAB, GLOBEFISH and Institutions and Experts Register), and microfiche services (fee), are available.

Description of Collection: The ASFIS bibliographic database includes over 200,000 references dating back to 1975. The database is available on DIALOG, CAN/OLE (Canada), DOCOCEAN (France), DIMDI (Germany), and via TELENET, TYMNET, and EURONET. Over 5,500 journal titles are monitored for ASFA. Centres that input into ASFA are located in Canada, France, Germany, the U.S.A., Japan, the U.S.S.R., Mexico, Portugal, the United Kingdom, Thailand, and China. UNDP's Aquaculture Development and Coordination Programmes (ADCP) has linked its aquaculture information network to ASFIS.

Subjects: Science, technology, and management of all aquatic resources, socioeconomic and legal aspects, exploration, and exploitation. Collection, storage, analysis, and dissemination of information relating to the management and development of aquatic resources. Fish capture technology and handling, processing, marketing, and distribution of fish and fishery products. Management and development of the living resources of the marine and inland waters as well as aquaculture and environmental protection. Fishery policy and planning advice particularly within the context of the new legal regime of the oceans.

Publications: Publications in the ASFIS Reference series include:

ASFIS-1 *List of Periodicals Monitored for the ASFIS Bibliographic Database.*
ASFIS-2 *ASFIS Subject Categories and Scope Descriptions.*
ASFIS-3 *ASFIS Guidelines for Bibliographic Description.*
ASFIS-4 *ASFIS Abstracting Guidelines.*
ASFIS-5 *ASFIS Guidelines for Subject Categorization and Indexing.*
ASFIS-6 *ASFIS Thesaurus* (revised and expanded version in preparation).
ASFIS-7 *ASFIS Geographic Authority List.*
ASFIS-8 *ASFIS Taxonomic Authority List* (expanded version in preparation).
ASFIS-9 *ASFIS Database User Guide.*
ASFIS-10 *ASFIS Authority List for Corporate Names and Acronyms* (final version in preparation).
ASFIS-11 *ASFIS Magnetic Tape Specifications and Record Format.*

Other publications available are listed in the *FAO Fisheries Department List of Publications and Documents*; *FAO Documentation—Fisheries*; *FAO List of Documents*; and *FAO Books in Print.*

(b) Aquaculture Development and Coordination Programme (ADCP)

This programme has six regional centres, located respectively as follows:

National Inland Fisheries Institute (NIFI), Kasetsart University Campus, Bangkhen, Bangkok, Thailand

Freshwater Aquaculture Research and Training Institute (FARTI), Dhauli, Bhubaneswar, Orissa, India

Southeast Asia Fisheries Development Centre (SEAFDEC), Aquaculture Department, Tigbauan, Iloilo City, The Philippines

Asia-Pacific Regional Research and Training Centre for Integrated Fish Farming, Wuxi, Jiangsu Province, PRC

African Regional Aquaculture Centre, Port Harcourt, Nigeria

Latin American Regional Aquaculture Centre (CERCA), Pirassununga, Brazil

The four centres in Asia are members of the Network of Aquaculture Centres in Asia (NACA), part of the FAO/UNDP Regional Programme. The programme has its headquarters in Bangkok, Thailand.

Tel: 579-4705 Telex: UNDEVPRO Bangkok 82392 ESCAP TH
Cable: NACABK Bangkok 9

Users Policy: Open to public.

Services: AQUIS Technical Data System, which handles alphanumeric and numeric data based on published aquaculture reports and bibliographic information. Hewlett-Packard 3000 computer systems with MINISIS software have been installed in each of the regional centres in Asia and in centres in Brazil. Photocopying services are available as well as database searching services.

Description of Collection: Each of the regional centres has a library collection.

Subjects: Major focus on aquaculture; specific interests vary with regional centre: Thailand—*Clarias, Channa, Macrobrachium, Pangasius, Trichogaster, Puntius*, mollusc culture; India—carp culture; Philippines—shrimp and brackish-water finfish culture; China—integrated fish farming; Nigeria—tilapia culture; Brazil—culture of Amazon species: *Cichla*, *Colossoma*, *Pimelodus*, *Prochilodus*, and *Mugil*.

(c) *Marine Science Contents Tables*

In 1966, FAO with the support of UNESCO began the monthly publication of *Marine Science Contents Tables* (*MSCT*), which reproduces the contents pages of over 100 'core' journals of marine sciences. It is available free of charge.

(d) *Freshwater and Aquaculture Contents Tables*

Freshwater and Aquaculture Contents Tables (*FACT*) was first published in 1966 and contains the contents pages of fifty 'core' journals. It is available monthly at no cost.

(e) GLOBEFISH

For several years marketing information services for the dissemination of fish marketing indicators for the most important fishery commodities, and the identification of medium-term trends in the marketing of these commodities, have been under development in Panama (INFOPESCA) and Kuala Lumpur (INFOFISH). Similar regional services are now being established in West Africa

(INFOPECHE) and the Arab countries (INFOSAMAK). Through these projects a network of users has been built up that covers virtually all the major fish-producing or fish-consuming nations. To take advantage of modern data communications, the information base is now being computerized and consolidated, as GLOBEFISH. It is regularly updated and is accessible worldwide using telecommunications networks. Agencies and institutions in developed countries will have on-line access to the database on subscription, which will defray the costs of operation. For developing countries without on-line access, the database will be used by the regional centres to generate marketing bulletins relevant to the region, and to run custom searches for the more specific needs of their users.

INFOFISH (P.O. Box 899, Kuala Lumpur 01-02, Malaysia. Tel: 914466) was established as a fisheries information organization handling fish and fish product information from Asia and the Pacific. It acts as a clearing house for market information on fish and fish products, exporters and importers. INFOFISH offers: (i) Trade Promotion Service (maintains the INFOFISH database, which is a register of fishery products in the Asia/Pacific region, producers/exporters of these products in the region, and importers of the products worldwide; publishes price information and market trends biweekly); (ii) a Technical Advisory Service with a library of fishery industry literature, which organizes seminars and training courses, publishes monographs and the bimonthly magazine, *INFOFISH Marketing Digest*, and special 'fact sheets'; and (iii) Market Information and Communication Service (publishes the *INFOFISH Marketing Digest*). INFOFISH maintains National Liaison Offices (NLO) in each member country to collect and disseminate information nationally.

INFOPESCA (Estafeta El Dorado, Apartado Postal 6-2400, Panama, Republic of Panama. Tel: 69-3477; Telex: 368460—INFOPESCA) has the same objectives as INFOFISH but was designed to serve users of fishery market information from Latin America.

(f) Indo-Pacific Fishery Commission (IPFC) (FAO, Maliwan Mansion, Phra Athit Road, Bangkok 10200, Thailand. Tel: 281-7844)

Maintains the regional fisheries library for FAO for Asia and the Pacific.

Users Policy: Open to public.

Services Publications are distributed irregularly. The public may use the library with permission from the Secretariat.

Description of Collection: 14,000 titles in agriculture, fisheries, aquaculture and forestry; 300 periodicals; maps; pamphlets; microfiche.

Subjects: Marine sciences; fisheries and aquaculture in Asia and the Pacific.

Publications: Proceedings of IPFC meetings, which are held every two years. Occasional papers, reports published after every IPFC seminar, workshop and training course.

(g) Bay of Bengal Programme (BOBP) (Post Bag No. 1054, Madras 600019, Madras, India)

BOBP is executed by FAO and financed by the Swedish International Development Authority. It disseminates technical literature on small-scale fisheries, and coastal aquaculture.

Publications: Technical publications and a newsletter.

Other International Organizations

474 International Association of Marine Science Libraries and Information Centres (IAMSLIC)

IAMSLIC is a forum for coordinating resource sharing within marine-related fields including aquaculture. Formally approved in 1979, the organization has over 100 members from seventeen different countries concerned with providing library and information services for marine science disciplines.

Services: In 1985 a programme for providing training in marine sciences for librarians and information scientists from developing countries was instituted.

Description of Collection: No collection maintained by IAMSLIC. Collections maintained by individual members.

Subjects: All disciplines related to marine sciences including aquaculture.

Publications: *Directory of Marine Sciences Libraries and Information Centres* and *Supplement*.

475 International Center for Living Aquatic Resources Management (ICLARM), Bloomingdale Building, Salcedo Street, Legaspi Village, Metro Manila, The Philippines

Tel: 818-0466 Telex: ITT 45658 ICLARM PM

Contact: Mr. Jay L. MacLean, Director, Information Services

International and regional information service.

Users Policy: Open to public.

Services: Fisheries and aquaculture library, photocopying (fee), searches of commercial databases (ASFA, BIOSIS, etc. at cost) (which can be accessed by visiting or writing to ICLARM's Selective Fisheries Information Service). SDI services, audiovisual equipment, and microfiche services are also available. ICLARM's information specialists will assist other institutions to develop their information services.

Description of Collection: 8,000 books and monographs, 600 serials.

Subjects: Fisheries; integrated farming; aquaculture; fisheries economics; stock assessment and management in tropical developing countries.

Publications: Acquisitions list. An excellent newsletter (subscription fee) relevant to topics in Southeast Asia and in tropical countries around the world. Technical publications (at cost): conference proceedings; studies and reviews; technical reports; bibliographies; and translations.

476 International Council for the Exploration of the Sea (ICES), Palaegade 2–4, DK-1261 Copenhagen, Denmark

Tel: (0) 1154225 Telex: 22498 ICES DK Cable: Merexploration, Copenhague

Contact: Mr. Edgar M. Thomasson, Librarian and Information Officer

Library collection for ICES.

Users Policy: Primarily set up for in-house use.

Services: Photocopying and interlibrary loans available to outside users, photocopying (fee).

Description of Collection: 10,000 journals; 300 books dating back to 1902.

Subjects: Oceanographic and fisheries science, including mariculture, in the North Atlantic and adjacent seas.
Publications: ICES reports and publications.

477 United Nations Environmental Programme (UNEP), UNEP Geneva (for Regional Seas or Publications), Palais des Nations, CH-1211 Geneva 10, Switzerland
Tel: (22) 985850 Telex: 28877 UNEP
Contact: Director RS/PAC

UNEP Headquarters (to access INFOTERRA): Box 30552, Nairobi, Kenya
Tel: 333930 Telex: 22068 UNEP
UNEP has two major information programmes that are relevant to marine science, INFOTERRA and the Regional Seas Programme.

(a) INFOTERRA

Environmental database that has over 100 input centres around the world. Over 70 per cent of the information in the database is from developing countries.
Users Policy: Access from office in Nairobi.
Services: Searches are conducted free of charge in Geneva. Users can send requests by letter, telephone, or telex. Output is provided in four languages. A printed directory of input centres is published every two years.
Description of Collection: Contains information from input centres located in about eighty countries.
Subjects: All aspects of environmental information including marine pollution and ocean dumping, and subjects related to aquaculture development.
Publication: Directory.

(b) Regional Seas Programme

Programmes for environmental management of coastal zones in ten defined areas of the world.
Users Policy: Open to public.
Services: Provide publications.
Subjects: Legal aspects, assessment, and management of marine areas in tropical regions.
Publications: Bibliographies, directories, surveys, data files.

10.3 Regional organizations

478 Asian Institute of Technology (AIT), Library and Regional Documentation Center, AIT, P.O. Box 2754, Bangkok 10501, Thailand
Tel: 523-9380-09
Contact: Ms. Daruna Somboonkun, Associate Director, Library
Research library and information centres.
Users Policy: Open to public.
Services: Research library; Environmental Sanitation Information Center (ENSIC) (provides information on integrated agriculture and waste recycling systems); bibliographies; searches of the inhouse databases and *Chemical Abstracts*, *Science Citation Index* and AGRIS; SDI services available by subscription.

Description of Collection: 130,000 volumes; 1,300 volumes on fisheries, aquaculture and related subjects; 20 fisheries and aquatic sciences journal titles of a total of 800 titles in the library.

Subjects: Aquaculture; aquaculture engineering; waste recycling systems; food engineering.

Publications: Publications of the information centres.

479 Southeast Asian Ministers of Education Organization (SEAMEO)
One SEAMEO Centre, BIOTROP—the Regional Center for Tropical Biology—offers information services in marine sciences.

BIOTROP, Bogor, Indonesia

Users Policy: Open to public.

Services: Clearing house for the storage and exchange of Southeast Asia tropical biological information.

Description of Collection: 7,000 books; 800 serials.

Subjects: Tropical biology including marine sciences; fisheries and aquaculture.

Publications: BIOTROP (quarterly newsletter); fisheries and biology manuals.

480 Southeast Asian Fisheries Development Center, Olympia Building, 4th Floor, 956 Rama IV Road, Bangkok, Thailand 10500
Tel: 235-2070 Telex: SUIDEV 20768TH Cable: SEAFDEC Bangkok

Three libraries located respectively in the Philippines, Singapore, and Thailand, as described below. Coordinates SAFIS and SEAFIS from Bangkok. Input centre for ASFIS (see [473C]).

(a) Secretariat's Information Programmes (address as shown above)
Contact: Information/Program Officer, and Project Leader of SAFIS and SEAFIS.

Coordinates the SAFIS and SEAFIS programmes and distributes the *SEAFDEC Newsletter* and other publications.

Publications: *SEAFDEC Newsletter* (quarterly compilation of information from all three departments, free if distributed by surface mail, charge for airmail service); a list of other SEAFDEC publications can be obtained by writing to the Information Officer. One of note is *Organizations/Agencies Concerned with Fisheries Development in Southeast Asia* (1984). This listing describes the fisheries services of thirty-four organizations/agencies in the region.

The Southeast Asian Fisheries Information Service (SAFIS) Programme, begun in 1982 and supported by the International Development Research Centre of Canada (IDRC), provides the following services to users in Southeast Asia and worldwide: (i) collection and classification of fishery extension manuals, brochures, pamphlets, and related aids; (ii) translation of selected literature from English to local dialects in Southeast Asia and vice versa for distribution to fisheries extension personnel in Southeast Asia; and (iii) preparation of new manuals, as requested. This programme provides excellent services, particularly to countries developing or establishing fisheries extension services in Asia. A list of manuals and translations published under the SAFIS programme is available from the Information/Program Officer; these include the acquisitions list and *Directory of Scientists and Technologists in Southeast Asia*.

The Southeast Asian Fisheries Information System (SEAFIS) began in 1984, supported by IDRC. The objectives of this project are: (i) to provide a facility that can act as an impetus to decision-making and planning for fisheries development in the region and which would, in turn, accelerate the pace of economic development in Southeast Asia; (ii) to act as a regional focal point and make available, in a centralized location, data and pertinent information on fisheries and related disciplines or technologies—specifically, to collect, store, analyse, and disseminate such information; (iii) to serve as the Southeast Asian regional input centre for the FAO Aquatic Science and Fisheries Information System (ASFIS) in Rome; and (iv) to provide access to important current regional fisheries information (both published and unpublished). SEAFIS's worldwide information services and its library services are available to the public.

(b) Marine Fisheries Training Department (address as above)

Users Policy: Open to public.

Services: Available to external users: current awareness, literature searches, photocopying (U.S. $0.15 per page), ILL, newspaper clipping service, numerical data service for fisheries statistics in the South China Sea, and an audiovisual service.

Description of Collection: Fisheries and aquaculture library.

Subjects: Applied fisheries research and training. Stock assessment and fishing techniques. Fisheries information.

Publications: Approximately 150 publications in the following series: Text and Reference Series; Miscellaneous Paper Series; Cruise Reports.

(c) SEAFDEC Aquaculture Department, P.O. Box 256, Iloilo City, The Philippines.

Contact: Mr. Emilio B. Gapit

Library of the Aquaculture Department of SEAFDEC and the Brackishwater Aquaculture Information System (BRAIS).

Users Policy: Open to public.

Services: Library, bibliographies, serials holdings, document delivery, training programmes in information sciences, SDI services, provided by SEAFDEC's Aquaculture Scientific Literature Service (ASLS).

The BRAIS project was initiated in 1984. It is also funded by IDRC. Input centres to BRAIS are located in each of the SEAFDEC member countries, namely Malaysia, the Philippines, Thailand, Singapore. Indonesia and India will be included in the future.

The services of BRAIS include: (i) Collection of documents on brackish-water aquaculture from the member countries, (ii) document storage and retrieval system, and (iii) bibliographic searches for libraries, researchers, technologists, and development personnel. It also has established a question-and-answer service for fish farmers and other information seekers, provides document delivery services, and maintains an up-to-date directory of researchers and research centres.

BRAIS circulates a quarterly current awareness newsletter, publishes state-of-the-art reviews, and issues special bibliographies.

The BRAIS system cooperates through the SEAFDEC Secretariat with international and regional information programmes such as the Aquatic Sciences and Fisheries Information System (ASFIS), the Aquaculture Information System (AQUIS), and the Agricultural Information Bank for Asia (AIBA).

Description of Collection: 6,000 monographs; 5,000 reprints; 2,500 pamphlets; 1,800 bound journals; 1,000 SEAFDEC publications; 320 microfiche titles; 30 rolls of microfilm; receives 420 serials titles regularly.

Subjects: Tropical aquaculture; aquaculture engineering; socioeconomics; fish biology; diseases; pests; feeds; nutrition, etc.

Publications: Extension manuals; annual reports; *Fishfarm News*; handbooks, technical reports, proceedings, quarterly research reports, serials holdings; acquisitions list (monthly); *Aquaculture Abstracts* (quarterly); *Milkfish Bibliography*.

(d) Marine Fisheries Research Department (MFRD), Changi Fisheries Complex, Changi Point, Singapore, Republic of Singapore
Tel: 5451625 Cable: SEAFDEC Singapore
Contact: Ms. Sundusia Rosdi, Librarian

Post-harvest technology library of SEAFDEC.

Users Policy: Access services through SAFIS and SEAFIS.

Services: Current awareness services, photocopying, interlibrary loans, and Table of Contents services available. Access to SAFIS and SEAFIS services. Databases searches can be conducted at the Singapore Institute of Standards and Industrial Research (SISIR) (costs vary for each search).

Description of Collection: Books, monographs, series, etc.

Subjects: Fisheries post-harvest technology (handling, preservation, and quality control of fish and fish products), and by-catch.

Publications: Marine Fisheries Research Department Technical Papers; accessions list.

481 South Pacific Commission (SPC), B.P.D. 5, Noumea Cede, New Caledonia
Tel: 26-20-00 Telex: 139 NM SOPACOM Cable: SouthPacom Noumea
Contact: B. M. Flores, Librarian

Has twenty member states, its secretariat being located in Noumea, New Caledonia.

Users Policy: Services available to public.

Services: Publications, translation to local languages, preparing bibliographies, numerical data service, database searching, ILL, microfiche services.

Subjects: Stock assessment; fishing techniques; aquaculture; fish processing; fisheries training; rural development; marine resources; deepsea fishing; tuna and billfish.

Publications: *SPC Newsletter* (quarterly, French and English); technical and scientific reports; fisheries handbooks. All publications and information services can be obtained by sending requests to the Commission's Noumea Office and from the Australian Branch Office, SPC Publications Bureau, Sydney South, N.S.W. 2000, Australia. *SPC Fisheries Newsletter* (quarterly); *Monthly News of Activities*; *Report of the South Pacific Conference*; annual report; accessions list.

11 NATIONAL ORGANIZATIONS

Argentina

482 Instituto Nacional de Investigación y Desarrollo Pesquero (INIDEP), Casilla de Correo 175, 7600 Mar del Plata
Tel: 51-7818 Telex: 39975 INIDP Cable: INIDEP
Contact: Head Librarian

Users Policy: Open to the public. Loans only to INIDEP staff. Special inter-library loans to the scientific community.

Services Available to Public: Photocopying. Reference restricted to day-loan for general public for photocopying outside the building. Photocopying services for requests through other libraries. Standard reference services. Bibliographies on demand.

Description of Collection: 2,400 books (monographs, expeditions, theses, seminars), 8,200 reprints, 900 serial titles (280 regular exchange and subscription titles, 150 maps.

Subjects: Coastal, pelagic, demersal, and freshwater fisheries resources development, marine ecology, aquaculture, fisheries oceanography, marine microbiology, food technology, ichthyology.

Publications: Irregular—on exchange: *Revista de Investigación y Desarrollo Pesquero, Contribuciones INIDEP, Memoria Anual.* Restricted distribution: *Boletín Informativo. Indice de Trabajos Publicados IBM-INIDEP 1962-1983, Lista de Publicaciones Periódicas Existentes en Biblioteca INIDEP.*

Brazil

483 Instituto de Pesca, 455 Avenida Francisco Matarazzo, 05001 São Paulo
Tel: (011) 62.01.91, (011) 62.28.78
Contact: Dr. Shitiro Tanji, Director of Marine Fisheries Division

Working Languages: Portuguese, English, Spanish, French.

Users Policy: Open to public.

Services Available to the Public: Library; abstract services; ILL; preparation of bibliographies (fee); numerical data services. Inputs into International Commission for Conservation of Atlantic Tuna, ASFIS, AGRIS.

Description of Collection: 2,100 books, 819 titles of periodicals.

Subjects: Fisheries biology, aquaculture, population dynamics, fishing technology, assessment and control of marine fisheries, conservation policy.

Publications: *Boletim do Instituto de Pesca.* ISSN 0046-9939. Irregular publication, with a minimum frequency of one issue/year. Current cost: Cr $10,000 ($1.50).

Canada

484 Fisheries and Oceans Canada, Pacific Biological Station Library, Nanaimo, British Columbia V9R 5K6

Tel: (604) 758-5202
Contact: Gordon Miller, Librarian

Users Policy: On-site use of collections—no restrictions; loans to outside users made on individual basis.

Services Available to Public: Interlibrary loans, on-line bibliography, current awareness services, bibliography.

Description of Collection: 120,000 volumes; 2,900 serials; 3,000 microforms.

Subjects: Salmon, marine biology, aquaculture, West Coast fisheries.

485 Fisheries and Oceans Canada, Scientific Information and Publications Branch (SIPB/DIPS), 12th Floor, 200 Kent Street, Ottawa, Ontario K1A 0E6
Contact: Director

Working Languages: English, French.

Users Policy: Publications available.

Services Available to Public: Publications.

Description of Collection: Not applicable.

Subjects: Marine sciences, fisheries, aquaculture, oceanography, pollution. ASFIS input centre for Canada.

Publications: Extensive list of publications including the *Canadian Journal of Fisheries and Aquatic Sciences* (monthly). Other irregular publications include: *Fisheries and Aquatic Sciences Manuscript Reports*; technical reports; special publications; bulletins; data reports; *Contractor Reports*; *Data Reports of Hydrography and Ocean Sciences*; *Industry Reports*; and translations of fisheries research from non-English/French-language journals. This and other publications are available as follows:

Priced publications are available from the Canadian Government Publishing Centre, Supply and Services Canada, Ottawa, Canada K1A 0S9.

Series are available in microfiche or photocopy from Micromedia Limited, 165 Hôtel de Ville, Hull, Quebec, Canada J8X 3X2.

ASFA is available in Canada on CAN/OLE, CISTI, Ottawa, Ontario, Canada K1A 0S2, and on the DIALOG system available through Micromedia Limited.

486 Fisheries and Oceans Canada, Bedford Institute of Oceanography Library, P.O. Box 1006, Dartmouth, Nova Scotia B2Y 4A2
Tel: (902) 426-3683 Telex: 019-31552
Contact: Head, Library Services

Users Policy: Open to public.

Services Available to Public: Reference services; on-line searching, interlibrary loans. Primarily for staff, to outside libraries and individuals by special arrangement. Xerox 4000; microform reader/printers. Library has computer-produced serials holdings list and KWIC (keyword in context) index to serial titles and to miscellaneous documents collection. Both may be available to other libraries in future on COM (computer output microfiche).

Description of Collection: 15,000 monographs; 17,000 volumes of periodical serials (including 3,700 volumes on cartridge microfilm); 20,000 non-periodical serial items (i.e. reports, etc., approximately 10,000 of which are on microfiche); 700 subscriptions to periodical serials.

Subjects: Chemical, physical, and environmental oceanography, marine geology and geophysics, marine ecology and pollution, fisheries, aquaculture, oceanography, hydrographic surveys and charting, marine technology. The emphasis is on the Canadian Atlantic seaboard, including the Gulf of St. Lawrence and the Eastern Arctic. Special collection of older material dealing with exploration and marine research in the Canadian Arctic.

Publications: *Bedford Institute of Oceanography Biennial Review*; technical reports.

487 Fisheries and Oceans Canada, Maritimes Regional Library, P.O. Box 550, Halifax, Nova Scotia B3J 2S7
Tel: (902) 426-6266 Telex: 3566
Contact: Regional Librarian

Users Policy: Primarily a government service with services to outside users with some restrictions.

Services Available to Public: Reference, circulation, on-line searching, SDI, interlibrary loans.

Description of Collection: 35,000 volumes; 1,350 serial titles; 50,000 microfiche titles.

Subjects: Fisheries, food technology, aquaculture, marine pollution, rare book collection of fisheries materials, Atlantic salmon.

Publications: Atlantic salmon bibliography.

488 Fisheries and Oceans Canada, Library, 240 Sparks Street, 8 West, Ottawa, Ontario K1A 0E6
Tel: (613) 995-9991 Telex: 053-4228 MARSI OTT
Contact: Librarian

Users Policy: Services to outside users.

Services Available to Public: Reference, on-line service, general inquiries, referrals, interlibrary loans.

Description of Collection: Approximately 20,000 volumes; currently 830 serial titles; approximately 3,000 items on microfiche.

Subjects: Fisheries, aquaculture, marine sciences, oceanography, and limnology.

Publications: Monthly acquisitions list; periodicals list (irregular).

489 International Development Research Centre (IDRC) Library, P.O. Box 8500, 60 Queen Street, Ottawa, Ontario K1G 3H9
Tel: (613) 236-6163 Telex: 053-3753 Recentre Ottawa. Envoy 100-IDRC LIBRARY

Users Policy: Open to the public. Special restrictions on use of confidential documents; ISRC confidential reports.

Services Available to Public: Photocopying, interlibrary loans, reference. On-line search services: MINISIS (FAO, BIBLIOL, DEVSIS).

Description of Collection: 40,000 books and documents, 5,000 serials titles.

Subjects: The collection is a current one relating to the economic and social development of the Third World, particularly its rural areas. In fisheries, the collection focuses on aquaculture, particularly in Southeast Asia, but also Latin America, the Caribbean and Africa. The collection reflects the research interests of IDRC-supported projects.

Publications: *Ex Libris*, IDRC reports, IDRC publications.

Chile

490 Instituto de Forrento Pesquera (Fisheries Development Institute), Avenida José Domingo Cañas 227, Casilla 1287, Santiago de Chile
Tel: 2239627 Cable: IFOP 3520001 ITT BOOTH SANTIAGO CHILE
Contact: Bibliotecaria
Working Language: Spanish.
Users Policy: Open to public.
Services Available to Public: Interlibrary loans.
Description of Collection: 7,000 books.
Subjects: Fisheries, oceanography, aquaculture, fisheries technology.
Publications: *Revista Científica Investigación Pesquera* (ISSN 0716-1328. Annual. U.S. $10.00); *Folleto Estudios y Proyectos de Desarrollo Pesquero* (Fortnightly. Distributed free of charge).

491 Instituto de Oceanología (Montemar), Universidad de Valparaíso, Avenida Barros Borgoño s/n, Montemar, Casilla 13-D, Viña del Mar, Valparaíso
Tel: 970420
Contact: Librarian
Working Languages: Spanish.
Users Policy: Open to public.
Services Available to the Public: Photocopying (fee).
Description of Collection: Not available.
Subjects: Marine biology and oceanography.
Publications: *Revista de Biología Marina* (ISSN 0080-2115, 2/yr., U.S. $4.00 (Amer.), U.S. $5.00 (Europe). Write to: Director, Instituto de Oceanología, etc. for exchange with similar publications); Publicaciones Ocasionales—irregular.

People's Republic of China

492 Institute of Marine Scientific and Technological Information, State Administration of the Sea of the People's Republic of China, 118 Qi Wei Road, Hedong District, P.O. Box 74, Tianjin, Tianjin
Tel: 44161 Cable: TIANJIN 5060
Contact: Mr. Li Shi Jie, Director of the Library
Input centre into ASFIS, and the Chinese National Oceanographic Data Centre.
Working Language: Chinese.
Users Policy: Open to public.
Services Available to Public: Library.
Description of Collection: Not available.
Subjects: Marine sciences; aquaculture and fisheries; oceanography.
Publications: *Marine Literature Bulletin, Oceanic Abstracts, China's Marine Weekly Paper, CNODC News* (Chinese National Oceanographic Data Centre), *Tide Tables*

for Chinese and Foreign Ports, annual reports, *International Marine Science Newsletter* (Chinese edition).

493 South China Sea Fisheries Research Institute, Chinese Academy of Fisheries Science, 219 Xingang Xilu, Guangzhou, Guangdong
Tel: 46320 Cable: 2389
Contact: Librarian
Working Languages: Chinese, Japanese, Russian, English, French.
Users Policy: Open to public.
Services Available to Public: Translations into English, French, Japanese, Chinese (fee); photocopying.
Description of Collection: 40,000 books.
Subjects: Artificial fish reefs; fishery resources management; otter board design; marine products preservation; artificial feed; nutrition; comprehensive utilization of marine products; effect on marine fisheries of oil and pesticides; aquaculture.
Publications: Research reports on fisheries.

Colombia

494 Instituto de Investigaciones Marinas de Punta de Betín (INVEMAR), P.O. Box AA 1016, Santa Marta, Magdalena
Tel: 35410 Contact: Director
Research library.
Working Languages: Spanish, English.
Users Policy: Open to public.
Services Available to the Public: Preparation of relevant publications, photocopying (fee).
Description of Collection: 1,835 volumes; 300 serial titles; 2,800 reports; 80 nautical charts.
Subjects: Fisheries biology, aquaculture, marine ecology, estuaries, marine flora and fauna, marine resources.
Publications: *Anales de Instituto de Investigaciones Marinas de Punta de Betín* (ISSN 0120-3959; annual).

Costa Rica

495 Sistema de Información en Ciencias Marinas, Recursos Pesqueros y Aquicultura (SIMPA), Apdo Postal 10318, 1000 San José
Tel: 24-41-72 Telex: 3338 Coni, C.R. Cable: CONICIT
Contact: Coordinator
Working Languages: Spanish, English.
Users Policy: Open to public.
Services Available to the Public: Photocopying, bibliographic reference service.
Subjects: Fisheries, aquaculture, fisheries resources, marine flora and fauna.
Publications: Lists of publications available in SIMPA's participating institutions.

Dominican Republic

496 Centro de Investigaciones de Biología Marina (CIBIMA), Jonas E. Salk 56, P.O. 1385, Santo Domingo, Distrito Nacional 7
Tel: 685-6682
Contact: Librarian
Working Languages: English, Spanish.
Users Policy: Publications only.
Description of Collection: Not available.
Subjects: Marine biology; mangroves and other coastal ecosystems; aquaculture.
Publications: Contribuciones (occasional papers) (10 per year). Mimeo, R.D. $1.00/copy.

Fiji

497 University of the South Pacific Library, P.O. Box 1168, Suva
Tel: 313900
Contact: Harold Holdsworth, Librarian
Users Policy: Full borrowing privileges, except periodicals, reference and bibliography, and the Pacific Collection.
Services Available to Public: Bibliographic (printed or cyclostyled). The library has satellite communication with eleven island countries, and Honolulu, Wellington, Sydney, Santa Cruz, and Vancouver. Interlibrary loans.
Description of Collection: 104,000 book volumes; 76,000 periodical volumes.
Subjects: Natural sciences, including aquaculture; education; social science. Pacific Collection (12,000 volumes).
Publications: *Pacific Collection Accession List* (quarterly), Selected Bibliography Series, *Bibliography of Periodical Articles Relating to the South Pacific, Pacific Union List of Periodicals*.

Finland

498 Finnish Game and Fisheries Research Institute, Fisheries Division, Korkeavuorenkatu 21, PB 193, SF-00131 Helsinki
Tel: 358-0-630794
Contact: Tuula Toivio-Syrjanen, Librarian
Users Policy: Open to public.
Services Available to Public: Photocopying (fee), interlibrary loans.
Description of Collection: 20,000 volumes, 400 current periodicals.
Subjects: Fisheries biology, fisheries economics, fisheries technology, aquaculture.
Publications: *Finnish Fisheries Research* (ISSN 0301-908X; irregular; on exchange basis); *Suomen Kalatalous* (ISSN 0085-6940; irregular; on exchange basis); *Meddelanden* (ISSN 0355-1652; irregular; on exchange basis); *Ministettuja Julkaisuja* (ISSN 0358-4623; irregular; on exchange basis). The publications *Finnish Fisheries Research, Suomen Kalatalous* and *Meddelanden* can be obtained from the Government Printing Centre, Marketing Division, P.O. Box 516, SF-00101 Helsinki 10, Finland.

France

499 Bureau National des Données Océaniques (BNDO), Institut Français de Recherche pour l'Exploitation de la Mer (IFREMER), IFREMER—Centre de Brest, B.P. 337, Cedex 29273 Brest
Tel: 98.45.80.55 Telex: OCEANEX 940627F
Contact: Head of the Bureau National des Données Océaniques (BNDO)
IODE and ASFIS input centre.
Working Languages: French, English.
Users Policy: Open to public.
Services Available to Public: Current awareness, literature searches, photocopying, interlibrary loans, numerical data services, database searches.
Description of Collection: 20,000 books, 1,800 periodical series.
Subjects: Oceanic exploration and exploitation (marine resources); marine technology; submarine technology; aquaculture; fisheries.
Publications: Accessions list; Annual Report on French Oceanic Cruises; IFREMER publications (cf. Raoul Piboub's report on that subject); on-line databases: ASFAdoc, PASCALo, OCEANIC, REVUMER, ROSCOPS, Geo IPOD, BIOCEAN, NEPTUNE, HYDRO (16 databases), AQUADOC, French Pollution Monitoring Network, BNDO Oceanic Bata Bank.

Federal Republic of Germany

500 Biologische Anstalt Helgoland (Biological Institute, Helgoland) Notkestrasse 31, D-2000 Hamburg
Tel: 040-89-69-30 Telex: 0214911
Working Languages: German, English.
Users Policy: Open to public.
Services Available to Public: Photocopying, interlibrary loans.
Description of Collection: 90,000 volumes.
Subjects: Marine zoology, marine botany, marine microbiology, experimental ecology, biological oceanography, marine pollution studies, aquaculture.
Publications: Accessions list; journal *Helgoländer Meeresuntersuchungen* (since 1894) (ISSN 0174-3597); Annual report: *Jahresbericht* (ISSN 0344-6573); book series: Diseases of Marine Organisms (ISBN 3-9800818-0-X and 3-9800818-3-4).

501 Bundesforschungsanstalt für Fischerei (Federal Research Centre for Fisheries), Palmaille 9, D-2000 Hamburg 50
Tel: (040) 38-90-5113 Telex: 02 15716 bfafi d
Contact: Dr. W. P. Kichner
Working Languages: German, English.
Users Policy: Open to public.
Services Available to Public: Current awareness (fee), literature searches (fee), photocopying (fee), preparation of bibliographies (fee), database searching (fee) (ASFA, FSTA, etc.). ASFIS input centre.
Description of Collection: 54,000 volumes, 950 current periodicals.
Subjects: Fisheries biology, fishery technology, fisheries statistics, marine biology, limnology, aquaculture, fish and shellfish processing.
Publications: *Archiv für Fishereiwissenschaft* (irregular, average 3 nos.

annually with occasional supplements, DM 80 to 100); *Informationen für die Fischwirtschaft* (quarterly; *c*. 220 p.; DM 40); *Informationen über die Fischwirtschaft des Auslandes* (9 per year; *c*. 400 p.; DM 18); *Jahresbericht Bundesforschungsanstalt für Fischerei* (free).

Ghana

502 Fishery Research Unit, P.O. Box B-62, Tema Ghana
Tel: 2346
Contact: Deputy Director of Fisheries
Working Language: English.
Users Policy: Open to public.
Services Available to Public: Numerical data services for fish catch statistics and hydrographic data.
Description of Collection: 534 books.
Subjects: Coastal fisheries of Ghana; fishing methods; hydrography; fisheries biology and statistics; aquaculture.
Publications: *Information Report*; *Marine Fishery Research Report*; biennial report on the fishery research.

India

503 Department of Aquatic Biology and Fisheries, University of Kerala, Beach P.O. Trivandrum 7, Kerala 695007
Tel: 71584/71138 Cable: AQUARIUM
Contact: Head
Working Languages: English, Malayalam.
Users Policy: Open to public.
Services Available to Public: Literature searches, photocopying, interlibrary loans.
Description of Collection: 6,000 volumes.
Subjects: Freshwater, estuarine, marine biology; oceanography; aquaculture.
Publications: *Journal of Aquatic Biology* (annual).

504 National Institute of Oceanography, Regional Centre, Pullepady Cross Road, P.B. No. 1913, Cochin, Kerala 682018
Tel: 30306 Cable: Oceanology Ernakulam India
Contact: Scientist-in-Charge
Working Language: English.
Users Policy: Open to public.
Services Available to Public: Interlibrary loans (fee), preparation of bibliographies (fee), zooplankton analysis (fee).
Description of Collection: 5,000 books, 10,000 reprints, 95 serial titles.
Subjects: Coastal zone management, aquaculture, planktonology, resource assessment, and aquatic pollution.
Publications: International Indian Ocean Plankton Atlases and Handbooks; Occasional Monographs. (*Mahasagar—Bulletin of the National Institute of Oceanography* is published from the headquarters at Dona Paula, Goa 403004, India.)

505 **University of Cochin, School of Marine Sciences**, Fine Arts Ave, Ernakulan, Cochin 682016, Kerala
Tel: 31957
Contact: Librarian
Working Language: English.
Users Policy: Open to public.
Description of Collection: 13,000 volumes.
Subjects: Studies on plankton; bottom fauna; bottom deposits; hydrography of Kerala coast; taxonomy, morphology, physiology, biochemistry, and ecology of fisheries, aquaculture.
Publications: Accessions lists; bibliographies.

Indonesia

506 **Lembaga Oseanologi Nasional** (National Institute of Oceanology), Jalan Pasir Putih No. l, P.O. Box 580/DAK, Jakarta 14420, DKI-JAYA
Tel: 683850 Cable: LONAS
Contact: Library and Documentation Center
Working Languages: Indonesian, English.
Users Policy: Open to public.
Services Available to Public: Publications. Inputs into SEAFIS and SAFIS.
Description of Collection: Not available.
Subjects: Indonesian waters; marine sciences: oceanography, microbiology, planktonology, ichthyology, geology, ecology, mariculture, pollution.
Publications: *Marine Research in Indonesia* (3 times per year); *Oceanologi di Indonesia* (*ODI*) (3 times per year, in English or Indonesian); *Oceanographic Cruise Report*; *Pewarta Oseana* (bimonthly); *Lonawarta* (quarterly newsletter); occasional publications, such as *Atlas Oseanologi Indonesia*.

Ireland

507 **Fisheries Research Centre**, Abbotstown, Castleknock, Dublin 15
Tel: 01-21011 Telex: 31236 FRCE1
Contact: Librarian
Working Languages: English, French, German.
Users Policy: Open to public.
Services Available to Public: Literature searches, photocopying, interlibrary loans, preparation of bibliographies and state-of-the-art reviews.
Description of Collection: 20,000 volumes.
Subjects: Stock assessment; fisheries; aquaculture; environmental impact studies; water pollution; fish pathology.
Publications: Accessions list; *Irish Fisheries Investigations*: Series A, *Freshwater*; Series B, *Marine*; *Fisheries Bulletin*; *Fisheries Research Centre Review*.

Italy

508 **Institute of Marine Environmental Sciences (ISAM)**, Via Balbi 5, I-16126 Genoa
Tel: 010/280955 Telex: 28114 Unistuge I

Working Languages: Italian, English, French, Spanish.
Users Policy: Open to public.
Services Available to Public: Photocopying (fee), preparation of bibliographies (fee).
Description of Collection: Not available.
Subjects: Biological and ecological sciences, marine fisheries, oceanography, limnology, physical sciences, pollution, demersal and pelagic fish, shrimps, other invertebrates, algae, plankton, benthos, aquaculture.
Publications: *Technical Report 1983* (no. 1), *1982* (no. 1), *1981* (no. 4), *1980* (no. 4); Monograph: *Seminari Internazionali sull'Inquinamento Marino* (1982), 151 p.; Monograph: *Il Mare e la Pesca* (1982), 160 p.

Jamaica

509 Discovery Bay Marine Laboratory, P.O. Box 35, Discovery Bay, St. Ann
Tel: 809-973-2241
Contact: Head
Working Language: English.
Users Policy: Open to public.
Services Available to Public: Photocopying.
Description of Collection: 5,000 volumes. Reprints: collection for Caribbean and tropical marine biology.
Subjects: Coral reef community ecology; coral physiology; echinoderm biology; marine lithification; biomechanics; ethology; aquaculture.
Publications: Not available.

Japan

510 Ministry of Agriculture, Forestry and Fisheries, Government of Japan, 2-1, 1-chome, Kasumigaseki, Chiyoda-ku, Tokyo 100.
Contact: Statistics and Information Department
Working Languages: Japanese, English
Services Available to Public: Publications, library.
Description of Collection: Information not available.
Subjects: Fisheries, aquaculture, post-harvest environment.
Publication: Fishery census of Japan.

Kenya

511 Department of Zoology, University of Nairobi, Chiromo Campus, P.O. Box 30197, Nairobi
Tel: 43181 Cable: Varsity Kenya
Working Language: English.
Users Policy: Open to public.
Services Available to Public: Assessment service.
Description of Collection: 10,000 volumes.
Subjects: Marine plankton ecology, fisheries and fish ecology, mariculture and marine pollution.
Publications: Bulletin of the Hydrobiological Society of East Africa, *Hydrobiologia Africana* (Quarterly. U.S.$10.00).

Korea

512 Korea Ocean Research and Development Institute (KORDI), 215-, Sucho-Dong, Kangnam-Gu, P.O. Box 17 Yeong Dong, Seoul 135
Tel: (562) 3051-5 Telex: KISTROK K27380 Cable: KISTROK SEOUL
Contact: Chief Librarian
Working Languages: Korean, English, French, Japanese.
Users Policy: Open to public.
Services Available to Public: Preparation of bibilographies, microfiche services.
Description of Collection: 12,000 books and reports.
Subjects: Physical oceanography; chemical oceanography; biological oceanography; geological oceanography; ocean engineering; marine policy; maritime and port affairs research; aquaculture.
Publications: *Bulletin of KORDI* (semi-annual; free); annual report (free); *Ocean Policy Newsletter* (monthly; free).

Kuwait

513 Agriculture Affairs and Fish Resources Authority, 21422 Safat
Tel: 439840 Telex: 30072 KT Cable: AGRIFISH
Working Languages: Arabic, English.
Users Policy: Information not available.
Services Available to Public: Publications.
Description of Collection: Information not available.
Subjects: Hydrology, hydrobiology, fish statistics, fish economics and management, aquaculture.
Publications: Principles for Protection of Fish Wealth and Organizing Fish Catching in Kuwait; The Sharks in the Arabian Gulf; Introduction to Study of Phytoplankton in the Arabian Gulf; Atlas of Hydrological and Fishery Data in the Territorial Waters of Kuwait; Fishes of Kuwait.

514 Kuwait Institute for Scientific Research (KISR), P.O. Box 24885 Safat, Kuwait City
Tel: 816988 Telex: KIST-KT 22299 Cable: Science Kuwait
Working Languages: Arabic, English.
Users Policy: Open to public.
Services Available to Public: Current awareness, literature services, photocopying, interlibrary loans (only within Kuwait), preparation of bibliographies and state-of-the-art reviews, database searching.
Description of Collection: 3,000 books; 250 serial titles; 100 technical reports; 5,000 reprints.
Subjects: Aquaculture; fisheries management; oceanography; marine biology.
Publications: Accessions list; *Kuwait Bulletin of Marine Science* (irregular; free to anyone who requests it; ISSN 0250-362X); International Technical Reports (departmental) on an *ad hoc* basis; *KISR Annual Research Report* (available free from the Publications and Public Information Division of KISR; ISSN 0225-4065).

515 Kuwait Institute for Scientific Research Marine Biology and Fisheries Library, (Branch of) National Scientific and Technical Information Center (NSTIC), P.O. Box 24885, Safat
Tel: (Kuwait) 618245 Telex: KISR KT 22299
Contact: Information Specialist
Users Policy: Open to the public; loan facilities available to KISR staff only.
Services Available to Public: Interlibrary loans: photocopies only, except to main NSTIC library. Reference service not restricted. On-line search services: DIALOG, DRS, SDC. Bibliographies on demand. Current awareness SDI.
Description of Collection: 3,000 book titles covering marine biology, oceanography, marine exology, aquaculture, fisheries, etc. (with direct access to NSTIC's 50,000 titles); 250 periodicals; 100 technical report series plus selected readings in microform; NTIS reports; reprint and pamphlet files; expedition reports; IOC depository; Arabian Gulf information file; nautical charts; departmental research reports.
Publications: Publications and reports of the Mariculture and Fisheries Department: *Kuwait Bulletin of Marine Science* (irregular; available free on request); International Technical Reports (irregular); KISR Annual Research Report (available free on request).

Madagascar

516 National Oceanographic Research Centre (CNRO), CNRO Nosy-be, P.O. Box 68, Nosy-be, Antsiranana 207
Tel: 613.73 Cable: OCEAN NOSY-BE
Contact: Chef du Service de Documentation Générale
Working Language: French.
Users Policy: Open to public.
Services Available to Public: Library, publications.
Description of Collection: 1,600 books, 1,000 serial titles.
Subjects: Fisheries; oceanography; ecology; pollution, aquaculture.
Publications: *Documents, Centre National de Recherches Océanographiques* (occasional papers); *Archives, Centre National de Recherches Océanographiques* (annual).

Malaysia

517 University of Agriculture Malaysia, Serdang, Selangor
Tel: 03-356101 Telex: MA 37454
Contact: Library Services
AGRIS input centre.
Working Languages: Bahasa Malaysia, English.
Users Policy: Open to public.
Services Available to Public: Current awareness, literature services, photocopying (fee), interlibrary loans (fee), table of contents (fee).
Description of Collection: 190,000 volumes.
Subjects: Chemical, physical and biological parameters in the South China Sea and Straits of Malacca; biodegradation of hydrocarbon in Malaysian waters; plankton studies in the South China Sea; coral reef ecology, management and conservation in Malaysian waters; turtle biology, management, and conservation; fish stock assessment; squid and cuttlefish biology; and aquaculture.

Publications: Accessions list. *Pertanika* (quarterly, MR$10.00 per issue); *Malaysian Agricultural Information Bulletin* (in Bahasa Malaysia) (quarterly, free).

Mexico

518 Centro de Información Científica y Humanística (CICH), Ciudad Universitaria, Apartado Postal 70–392 y 20–281, 04510 México, D.F.
Tel: 548-08-58 Telex: 17-74523 UNAMME
ASFIS input centre.
Working Languages: Spanish, English.
Users Policy: Open to public.
Services Available to Public: Reference library, photocopying, retrospective bibliographic searches, special searches, SDI; ASFA and Biblat databases.
Description of Collection: 1,800 journals; also has a collection of specialized dictionaries, encyclopedias, directories, catalogues and mimeographs.
Subjects: Oceanography, fisheries, aquaculture.
Publications: *CLASE* (quarterly index to 600 Latin American periodical titles in general science and humanities); *Periodica* (quarterly, index to over 700 Latin American journal titles in technology and hard sciences); *Bibliografía Latinoamericana I* and *II* (available on-line through QUESTEL in France; includes articles from and on Latin America appearing in foreign journals); *Portal* (monthly current awareness bulletin displaying table of contents of more than forty library and information science journals).

Mozambique

519 Instituto de Investigação Pesqueira (IIP), Av. Mao Tse Tung No. 387, P.O. Box 4603 Maputo
Tel: 74 41 33 Telex: 6-497 PEIXE-MO
Contact: Librarian.
Working Languages: Portuguese, French, Spanish, English.
Users Policy: Open to public.
Services Available to Public: Accessions list, literature searches.
Description of Collection: 2,000 books, 1,400 pamphlets, 80 serials, charts and maps.
Subjects: Fisheries biology; oceanography; limnology; fisheries technology application; fish product technology application; aquaculture; fish inspection and quality control.
Publications: *Revista de Investigação Pesqueira* (irregular); *Boletim de Divulgação* (irregular).

Netherlands Antilles

520 Foundation Carmabi, Piscaderabaai, P.O. 2090, Willemstad, Curaçao
Tel: 624242 Cable: CARMABI-CURACAO
Working Languages: English, Dutch.
Users Policy: Publications only.
Services Available to Public: Publications only.
Description of Collection: 1,630 volumes.

Subjects: Ecological sciences, resource management, coral ecosystems, mangrove ecosystems, research on commercial crustacea, aquaculture.

Publications: Carmabi Collected Papers; STINAPA Series; STINAPA Documentation Series.

New Caledonia

521 Centre Orstom de Noumea, ORSTOM, B.P.A 5, Noumea
Tel: 26-10-00 Telex: Orstom 193 NM Cable: Orstom Noumea
Contact: Librarian

Working Languages: French, English.

Users Policy: Open to public.

Services Available to Public: Interlibrary loans; table of contents service.

Description of Collection: 1,100 serial titles; 5,000 volumes.

Subjects: Fisheries; oceanography of the South Pacific; aquaculture.

Publications: *Recueil de Travaux* (ISSN 0078-2730; once a year; free); *Rapports Scientifiques et Techniques* (irregular, free).

New Zealand

522 Division of Marine and Freshwater Science, Department of Scientific and Industrial Research, Evans Bay Parade, Box 12-346, Wellington
Telex: 3276/Research Cable: OCEANGRAPH WELLINGTON
Input into SIRIS (printed version is *N.Z. Science Abstracts*).

Working Language: English.

Users Policy: Open to public.

Services Available to Public: Literature searches (fee), photocopying (fee), database searching (fee).

Description of Collection: 50,000 volumes.

Subjects: Marine geoscience, physical oceanography, marine biology, freshwater science, aquaculture.

Publications: N.Z. Oceanographic Institute Memoirs (ISSN 008-7903; irregular); NZOI Records (ISSN 0110-618X; irregular); NZOI Oceanographic Summaries (ISSN 0111-1302; irregular); NZOI Oceanographic Field Reports (ISSN 0110-5205; irregular); N.Z. Oceanographic Institute Miscellaneous Publications (ISSN 0510-0054; irregular); N.Z. Oceanographic Institute Hydrology Station Data (ISSN 0112-0345; irregular); N.Z. Oceanographic Institute Biology Station Data (ISSN 0112-3335; irregular); *Division of Marine and Freshwater Science Annual Report* (ISSN 0112-3238; annual); *Research in Oceanography and Freshwater* (ISSN 0112-4536; annual). Cost varies according to size of issue. Most available on exchange as well as subscription.

523 Fisheries Research Centre, Greta Point, Evans Bay Parade, P.O. Box 297, Wellington
Tel: 861-029
Contact: Scientific Officer

AGRIS input centre.

Working Language: English.

Users Policy: Open to public.

Services Available to Public: Current awareness, literature searches, photocopying (fee), interlibrary loans, preparation of bibliographies, database searching (fee) on DIALOG.

Description of Collection: 11,138 periodical volumes, 2,340 books.

Subjects: Marine and freshwater fisheries; aquaculture; electronics; acoustics; mathematical modelling of fisheries.

Publications: Fisheries Research Bulletins (ISSN 0110-1749; irregular; free); Occasional Publications (ISSN 0110-1765; irregular; free); Information Leaflet (ISSN 0110-4519; irregular; free); *Shellfisheries Newsletter* (ISSN 0112-501X; quarterly).

Nicaragua

524 Instituto Nicaragüense de la Pesca (INPESCA), Km 4 1/2 Carretera Sur, Frente la Embajada de Estados Unidis, Apartado Postal 2020, Managua

Tel: 61369 Ext 69 Telex: 1309 INPESCA

Contact: Director

Working Languages: Spanish, English.

Users Policy: Open to public.

Services Available to Public: Photocopying (fee), interlibrary loans.

Description of Collection: 2,651 books; 500 serial titles.

Subjects: Fishery biology, marine biology, fishery statistics and economics, fishery planning and aquaculture.

Publications: Accessions list; *Boletín Técnico Centro Investigaciones Pesqueras, INPESCA; Boletín Informativo INPESCA; Boletín Estadístico Pesquero. Ofic. Estadísticas INPESCA*; Annex: Reports published.

Nigeria

525 Department of Zoology, University of Ibadan, Hydrobiology and Fisheries Research Unit, Ibadan, Oyo

Tel: 400550 Cable: Unibadan, Nigeria

Working Language: English

Users Policy: Open to public.

Services Available to Public: Photocopying (fee), interlibrary loans, preparation of bibliographies, microfilm services.

Description of Collection: 6,000 volumes.

Subjects: Aquaculture; fish biology; aquatic pollution.

Publications: List not available.

Norway

526 Department of Marine Biology, University of Bergen, N-5065 Blomsterdalen

Tel: 05-22-62-00

Working Languages: Norwegian, English.

Users Policy: Only publications.

Services Available to Public: Publications.

Description of Collection: 6,000 books, 50,000 reprints, 350 serial titles.

Subjects: General marine ecology and biological oceanography; ecosystem ecology; study of pelagic systems; benthic ecology; population ecology on pelagic and benthic species; deep water ecology (Norwegian Sea); studies on the aquaculture potential of benthic organisms.

Publications: *Sarsia* (ISSN 0036-4827; 1961–; Cost: NOK 450 per year, postage included; 4 issues per year, *c*. 320 p. Subscribe directly to: Sarsia, Department of Marine Biology, N-5065 Blomsterdalen, Norway).

527 **Institute of Marine Research, Directorate of Fisheries**, Nordnesparken 2, P.O. Box 1870–72, N-5000 Bergen
Tel: 47-5-327760 Telex: 42297 Ocean N.
Contact: Librarian
Working Languages: Norwegian, English.
Users Policy: Open to public.
Services Available to Public: Literature searches, photocopying, interlibrary loans.
Description of Collection: 63,650 volumes.
Subjects: Oceanographic cruises; fisheries; marine biology; resource assessment; aquaculture.
Publications: *Årsmelding fra Fiskeridirektoratets Havforskningsinstitutt* (in the main serial publication: *Årsberetning vedkommende Norges Fiskerier*; annual); *Fisken og Havet* (irregular); Fiskeridirektoratets Skrifter, serie Havundersokelser (irregular). (Exchange agreements for literature on similar topics.)

Pakistan

528 **Zoological Survey Department, Karachi**, Block 61, Pakistan Secretariat, Shahrah-e-Iraq, Karachi I
Tel: 521363 Cable: 'ZOOLOGY', KARACHI
Contact: Director
Working Languages: English.
Users Policy: Open to public.
Services Available to Public: Library; interlibrary loans.
Description of Collection: 50,000 volumes.
Subjects: Research on the systematic, population, distribution, biology and biochemistry of marine fauna of Pakistan, aquaculture.
Publications: *Records/Zoological Survey of Pakistan* (free or with exchange).

Panama

529 **Dirección General de Recursos Marinos**, Edificio Loteria Piso 15, Ministerio de Comercio e Industrias, 3318 Panama City 4
Tel: 27-4211 Telex: 2256 Comerin PA
Working Languages: Spanish, English.
Users Policy: Open to public.
Services Available to Public: Translation (charge per page).
Description of Collection: 14,000 volumes.
Subjects: Marine sciences; statistics; pollution; fisheries; aquaculture.
Publications: Accessions list; *Estadísticas Pesqueras* (annual; free).

Peru

530 **Instituto del Mar du Perú**, Esq. Gamarra y Gral. Valle, S/N Chucuito, Callao, Aptdo. 22, Callao
Tel: 297630
Working Language: Spanish.
Users Policy: Open to public.
Services Available to Public: Photocopying, current awareness.
Description of Collection: 2,800 books, 7,500 reprints, 3,500 reference books.
Subjects: Marine and freshwater research, fisheries, oceanography, fishery engineering, aquaculture.
Publications: *Informe*; *Boletín* (irregular).

Philippines

531 **Bureau of Fisheries and Aquatic Resources (BFAR)**, 860 Arcadia Building, Quezon Avenue, Quezon City, P.O. Box 623, Manila 2801
Tel: 96-54-28 Telex: BFAR 2566 PU
Contact: Librarian
Working Languages: English, Filipino.
Users Policy: Open to public.
Services Available to Public: Current awareness, literature searches, photocopying (fee), preparation of bibliographies (fee), newspaper clipping service (fee), table of contents service (fee).
Description of Collection: 3,000 books and unbound periodicals and monographs; serials: 100 titles; annuals, proceedings, reports, magazines, and news clippings.
Subjects: Fisheries and aquatic sciences; aquaculture; oceanography; ecology; pollution; aquaculture technology; management and conservation; marketing; economics; statistics; laws and legislation; biology and zoology; microbiology; processing and utilization and related disciplines.
Publications: Accessions list; annual report (limited circulation); *Fisheries Newsletter* (quarterly—on exchange only); *Philippine Journal of Fisheries* (semi-annual; $50.00 including postage); *Philippine Fisheries Statistics* ($50.00); General Information Series and Technical Paper Series (irregular; free when copies are available).

532 **University of the Philippines in the Visayas, Library**, Iloilo City 5901
Tel: 7-92-48
Contact: Librarian
Users Policy: Open to the public. Use of collection upon accreditation.
Services Available to Public: Photocopying and reference upon accreditation. Bibliographic services and training programmes.
Description of Collection: The collection in the natural sciences, fisheries, aquaculture, marine sciences and related fields holds 6,500 books, 328 serial titles and 7,500 reprints.
Publications: *DANYAG: Journal of Studies in the Humanities, Education and the Sciences, Basic and Applied*; research monographs.

Portugal

533 Instituto Nacional de Investigação das Pescas (INIP) (National Institute of Fisheries Research), Av. Brasília, 1400 Lisbon
Tel: 610814 Telex: 15857 INIPP
Contact: Information Centre
Working Languages: Portuguese, English, French.
Users Policy: Open to public.
Services Available to Public: Photocopying, interlibrary loans.
Description of Collection: 7,080 books; 10,000 reprints; 1,728 serial titles.
Subjects: Fisheries; marine biology; limnology, aquatic pollution; fish technology, aquaculture.
Publications: Accessions list; *Boletim do Instituto Nacional de Investigação das Pescas; Publicações Avulsas do INIP; Relatorios do INIP* (irregular; available on exchange basis).

Puerto Rico

534 **University of Puerto Rico**, Mayaguez Campus Library, Marine Sciences Collection, Mayagüez 00708
Tel: (809) 832-4040
Contact: Head, Marine Sciences Collection
Users Policy: Open to the public. Users outside of the marine sciences department allowed to use the library for reference only. Theses not available for loan.
Services Available to Public: Photocopying, interlibrary loan, reference. On-line search services: DIALOG available through Mayaguez Campus Library, not restricted.
Description of Collection: 1,169 books (volumes), 879 periodical titles, 3,761 periodical volumes, 6,248 documents, 130 theses, 885 microforms, 10,026 reprints. Marine biology, fish biology, aquaculture, marine invertebrates, marine botany, marine ecology; chemical, physical, geological, and biological oceanography.
Publications: *Department of Marine Sciences Contributions* (annual; available on exchange basis).

Qatar

535 **Department of Marine Sciences, University of Qatar**, Box 2713, Doha
Tel: 873-910 Ex. 34 Telex: 4630 Unvsty-HD Cable: UNIV-QATAR
Contact: Director, Library
Working Languages: Arabic, English.
Users Policy: Open to public.
Services Available to Public: Literature services; abstract service for theses in Arabic; interlibrary loans for Arabian Gulf universities; table of contents service; database searching; microfiche services.
Description of Collection: Whole university: 129,000 volumes, 1,900 periodical subscriptions.
Subjects: Marine sciences, fishery sciences, aquaculture.
Publications: Accessions list; *University of Qatar Science Bulletin*.

Senegal

536 Centre de Recherches Océanographiques de Dakar-Thiaroye (CRODT), KM 10.5, Route de Rufisque, B.P. 2241, Dakar
Tel: 34-05-34
Contact: Librarian
Working Language: French.
Users Policy: Open to public.
Services Available to Public: Current awareness, photocopying (fee).
Description of Collection: Information not available.
Subjects: Marine sciences, fisheries resources, aquaculture.
Publications: Accessions list, 1967–74: Documents Scientifiques Provisoires (DSP) du CRODT (occasional papers); 1974–: (a) Documents Scientifiques (DS) du CRODT (occasional papers); (b) Archives du CRODT (internal reports).

537 Directeur de l'Océanographie et des Pêches Maritimes (DOPM), 1, rue Joris, B.P. 289, Dakar
Tel: 21-565-78
Contact: Documentaliste.
Working Language: French.
Users Policy: Open to public.
Services Available to Public: Literature searches, photocopying (fee).
Description of Collection: Information not available.
Subjects: Marine resources, fisheries technology, commercial fisheries, oceanography, aquaculture.
Publications: *Répertoire des Textes Législatifs*; acquisitions list; *Bibliographie des Études Rédigées au Niveau de la DOPM; Bulletin de Sommaire.*

Sierra Leone

538 Institute of Marine Biology and Oceanography, Fourah Bay College, University of Sierra Leone, Freetown
Tel: 50775 Cable: IMBO FOURAH BAY COLLEGE
Contact: Senior Research Fellow
Working Language: English.
Users Policy: Publications only.
Services Available to Public: Publications only.
Description of Collection: 500 volumes.
Subjects: Fisheries, oceanography, fishery biology and ecology, fish stock assessment, estuarine pollution, aquaculture.
Publications: *Bulletin of the Institute of Marine Biology and Oceanography* (annual; can be purchased directly from the Institute).

South Africa

539 Sea Fisheries Research Institute (of the Chief Directorate Marine Development of the Department of Environment Affairs of the Republic of South Africa), Martin Hammerschlag Way, Private Bag X2, Rogge Bay, Cape Town 8012
Tel: 012-211480 Telex: 526425 Cable: Plankton, Cape Town
Contact: Librarian

Users Policy: Open to public.

Services Available to Public: Photocopying (fee), interlibrary loans, numerical data centre, microfiche services.

Description of Collection: 4,000 books, 1,300 serial titles (700 currently received), 19,000 reprints, 600 charts.

Subjects: Physical, chemical, and biological oceanography; fisheries biology (marine macrophytes, invertebrates, fishes, guano-producing seabirds, seals); marine pollution (including mineral oil); aquaculture.

Publications: Annual Report (annual, separate English and Afrikaans issues); Investigational Report (irregular; monographs, English and Afrikaans, bilingual abstracts); *South African Journal of Marine Science* (irregular; collection of papers, bilingual abstracts); Special Report (irregular; monographs, not research reports, English or Afrikaans). All available on exchange gratis to related organizations worldwide. Write to Editor.

Thailand

540 Facility of Fisheries Library, Kasetsart University, 50 Paholyothin Street, Bangkok 1093
Tel: 5790113 Ext. 376
Contact: Head, Department of Marine Science

Working Languages: Thai, English.

Users Policy: Open to public.

Services Available to Public: Literature searches, photocopying (fee), interlibrary loans.

Description of Collection: 2,915 books, 366 serial titles.

Subjects: Biology, oceanography, fisheries, aquaculture.

Publications: Accessions list; *Kasetsart University Fishery Research Bulletin* (ISSN 0125-796X; annual; exchange); *Notes from Faculty of Fisheries, Kasetsart University* (ISSN 0125-7978; annual; exchange); texts on fisheries, most of them in the Thai language.

541 Phuket Marine Biological Center, P.O. Box 60, Phuket Province
Tel: 212357

Working Languages: Thai, English.

Users Policy: Open to public.

Services Available to Public: Photocopying.

Description of Collection: 1,500 books; 5,000 reports; 70 monographs; 50 periodical subscriptions.

Subjects: Marine fisheries, marine pollution, marine environment, coastal ecology, taxonomy, aquaria, aquaculture.

Publication: *Research Bulletin of the Phuket Marine Biological Center* (irregular; for exchange only).

Trinidad and Tobago

542 Institute of Marine Affairs (IMA), Hilltop Lane, P.O. Box 3160, Carenage
Tel: (809) 625-102 Ext. 491-3 Cable: MAR INAF
Contact: Manager, Information Science Division

Working Language: English.
Users Policy: Open to public.
Services Available to Public: Database searching.
Description of Collection: 1,770 books; 4,600 serial titles.
Subjects: Oceanography, fisheries biology; coastal area planning and management; geology; water quality; law of the sea/maritime law; marine education, aquaculture.
Publications: Annual report (free); *Research Report U.S.* ($10.00 as available).

Tunisia

543 Institut National Scientifique et Technique d'Océanographie et de Pêche (INSTOP), 28 rue de 2 Mars 1934, Salammbô, Tunis 2025
Tel: 01-27-56-32 Cable: INSTOP 2025 Salammbo Tunisie
Contact: Documentation Center.
Working Languages: Arabic, French.
Users Policy: Open to public.
Services Available to Public: Photocopying.
Description of Collection: Information not available.
Subjects: Aquaculture, marine resources, oceanography, marine pollution, fisheries techniques.
Publications: Bulletins, reports.

Turkey

544 Middle East Technical University, Institute of Marine Sciences, P.K. 28, Erdemli, Icel
Tel: 9-7585-1842 Telex: 67208 dms-tr
Contact: Librarian
Working Languages: English, Turkish.
Users Policy: Open to public.
Services Available to Public: Interlibrary loans.
Description of Collection: 1,400 books; 1,300 serial titles; 170 volumes of collected reprints.
Subjects: Physical oceanography, marine biology, marine chemistry, marine pollution, marine geology and geophysics, fisheries, aquaculture.
Publications: None.

United Kingdom

545 Marine Biological Association of the United Kingdom, The Laboratory, Citadel Hill, Plymouth, Devon PL1 2PB, England
Tel: (0752) 21761 Cable: Laboratory Plymouth
Contact: Head, Library and Information Services.
National input centre into ASFIS/ASFA. One of the most complete marine biological sciences collections in the world.
Users Policy: Open to public.
Services Available to Public: By contract: enquiry, information, special bibliographies, literature searching, data analyses; the Association coordinates a marine pollution information centre.

Subjects: Marine biology, fisheries, aquaculture, marine pollution, marine ecology.

Publications: *Journal of the Marine Biological Association* (since 1887; quarterly); bibliographies; *Marine Pollution Research Titles*. Publications list available.

546 Ministry of Agriculture, Fisheries and Food, Directorate of Fisheries Research, Fisheries Laboratory, Pakefield Road, Lowestoft, Suffolk NR33 0HT, England
Tel: Lowestoft 62244 Telex: 97470 Cable: FISHLAB Lowestoft
Contact: Librarian

ASFIS input centre. Supplies data to the U.K. Data Centre, Marine Information and Advisory Service, Brook Road, Wormley, Godalming, Surrey GU8 5UB, England.

Working Language: English.

Users Policy: Open to public.

Services Available to Public: Photocopying (fee), interlibrary loans.

Description of Collection: Over 1,000 current serial titles; several thousand textbooks, reports, reprints, pamphlets, translations, etc.

Subjects: Fisheries research (including shellfish research); aquatic pollution; aquaculture.

Publications: Information not available.

United States: South Carolina

547 U.S. Department of Commerce, NOAA, National Marine Fisheries Service, Southeast Fisheries Center, Charleston Laboratory, 217 Fort Johnson Road, P.O. Box 12607, Charleston, SC 29412
Tel: (803) 762-1200
Contact: Chief, Library and Technical Information Services

Users Policy: Open to public.

Services Available to Public: Interlibrary loans.

Description of Collection: 8,000 books, reports, bound journals; 125 serial subscriptions.

Subjects: Chemical and toxicological evaluation; composition; nutritional value and quality of fish and seafood products; aquaculture.

Publications: Items in NOAA Technical Report or Technical Memorandum series.

United States: National

548 U.S. National Marine Fisheries Service, Office of Science and Technology, NMFS (F/S), 3300 Whitehaven St., N.W., Washington, DC 20235
Tel: (202) 634-7469 Telex: 650 221
Contact: Deputy Assistant Administrator for Science and Technology, NMFS

Users Policy: Open to public.

Services Available to Public: Accessions list; current awareness service; translations; literature searches; abstracting service; photocopying;

interlibrary loans; bibliographies. Input into ASFIS and the FAO's GLOBEFISH.

Description of Collection: Numerous fishery centre libraries.

Subjects: Marine biology; ecosystems; mariculture; technology; marine resource economics; fishery product developments.

Publications: *Fishery Bulletin* (quarterly); *Marine Fisheries Review* (quarterly); Technical Reports (irregular); Technical Memoranda (irregular); *Fisheries of the U.S* (annual).

United States: California

549 Scripps Institution of Oceanography Library, University of California, San Diego C-075C, La Jolla, CA 02093
Tel: (714) 452-3274

Users Policy: Open to public, circulation limited.

Services Available to Public: Reference services; on-line search services (Lockheed DIALOG, SDC ORBIT, BRS); interlibrary loans. Photocopy machines available.

Description of Collection: 128,962 bound volumes; 6,995 (3,189 active, 3,766 inactive) serial titles; 34,543 maps; 193 reels microfilm; 1,008 microcards; 5,384 microfiche; 30 motion pictures; 31,435 documents/reports/translations; 20,611 reprints; 19 recordings; 18 photos; 130 linear feet of archives.

Subjects: Fisheries; marine sciences; oceanography; aquaculture.

Publications: SIO Library Serials List; SIO Contributions; *SIO Bulletin*; annual report; *Catalogs of the SIO Library* (published by G. K. Hall); KWIC Index to the Documents Collection.

United States: Florida

550 Rosenstiel School of Marine and Atmospheric Science, University of Miami, 4600 Rickenbacker Causeway, Miami, FL 33149
Tel: (305) 361-4007
Contact: Librarian

Users Policy: Limited services to those outside the local scientific community.

Services Available to Public: Interlibrary loans.

Description of Collection: 35,000 books and bound periodicals; 25,000 reprints; 2,100 microforms, 1,000 current serial titles; 45 sets of oceanographic expedition report series.

Subjects: All marine sciences, especially of tropical seas, marine biology and fisheries; geology and geophysics; physical and chemical oceanography; ocean engineering; atmospheric sciences; aquaculture.

551 U.S. Dept. of Commerce, NOAA, National Marine Fisheries Service, Panama City Laboratory Library, 3500 Delwood Beach Road, Panama City, FL 32407-7499
Tel: (904) 234-6541
Contact: Librarian

Users Policy: Open to public. Loans to the public with permission.

Services Available to Public: Photocopying, interlibrary loans, and reference.

DIALOG search service. On-line searching for staff only. Laboratory publications available upon request.

Description of Collection: 2,500 books, 400 periodical titles, 8,000 technical reports, and 7,800 reprints.

Subjects: Fisheries science, marine biology, and oceanography, aquaculture.

Publications: Annual list of publications available from the laboratory.

United States: Massachusetts

552 **Woods Hole Oceanographic Institution**, Woods Hole, MA 02543
Tel: (617) 548-1400
Contact: Research Librarian

Users Policy: Reference services available to outside users. On-site use only for Data Library Collection.

Services Available to Public: On-line search services (Lockheed DIALOG, SDC ORBIT, and DOE RECON). Microreproduction facilities and photocopying equipment available. Interlibrary loans.

Description of Collection: 20,000 technical reports; 700 atlases; 10,000 maps; 1,600,000 underwater photographs. Ships' libraries, small departmental collections, Institution archives.

Subjects: Oceanography, geology, geophysics, marine biology, ocean engineering, meteorology, marine policy, aquaculture.

Publications: Collected reprints; *Oceanographic Index* (edited by Mary Sears, published by G. K. Hall); *Document Library Monthly Accession List* and annual *Bibliography of Technical Reports*.

United States: Rhode Island

553 **University of Rhode Island, International Center for Marine Resource Development, Library**, Kingston, RI 02881
Tel: (401) 792-2938
Contact: Librarian

Users Policy: Open to the public. In-person and telephone inquiries welcome.

Services Available to Public: Standard reference services; specialized services to international agencies, government agencies, and libraries of developing countries; interlibrary loans.

Description of Collection: 4,000 books and documents; 200 journals and newsletters.

Subjects: Fisheries and aquaculture in developing countries; some related marine policy materials.

554 **Division of Marine Resources Library, University of Rhode Island**, Bay Campus, Narragansett, RI 02882
Tel: (401) 792-6211
Contact: Jane Miner, Librarian

Users Policy: Serves all members of the marine community.

Services Available to Public: Maintains files for Rhode Island Marine Bibliography; interlibrary loans.

Description of Collection: 3,000 volumes; 100 periodicals; technical reports, environmental impact statements, Northeast and pertinent national documents.

Subjects: Concentration on grey literature; coastal zone management; energy; fisheries management; coastal hazards and environmental protection; aquaculture.

Publications: Bimonthly acquisitions list; bulletins, newsletters, reports are distributed in cooperation with the Division of Marine Resources Publication Unit.

United States: National

555 National Sea Grant Depository, University of Rhode Island, Bay Campus, Narragansett, RI 02882
Tel: (401) 792-6114
Contact: Director

Users Policy: Thirty-day loans to any interested parties.

Services Available to Public: Cumulative and yearly indexes; monthly acquisitions listings; packaged literature searches and manual literature searches; interlibrary loans.

Description of Collection: 7,000 documents including technical reports, reprints, marine advisory reports, conference proceedings, etc.

Subjects: Marine resources, aquaculture.

Publications: *Cumulative Sea Grant Publications Index 1968–80* (microfiche); *Sea Grant Publications Index 1980*; monthly acquisitions lists.

United States: Virginia

556 Virginia Institute of Marine Science (VIMS), Library, Gloucester Point, VA 23062
Tel: (804) 642-2111
Contact: Librarian

Users Policy: Library open to public.

Services Available to Public: On-line searching (DIALOG and OASIS), photocopier, sale of VIMS publications; interlibrary loans.

Description of Collection: 24,000 bound volumes (monographs and serials); 1,500 serial titles including exchange material from foreign fisheries departments.

Subjects: Chesapeake Bay; marine biology and geology; oceanography; estuarine ecology and environmental sciences; sport fishing; aquaculture.

Publications: Acquisitions list; VIMS publications list; serials list; *Chesapeake Bay Bibliography*; *VIMS Contributions*.

United States: National

557 U.S. Department of Agriculture, National Agricultural Library, Aquaculture Information Center, Public Services Division, Room 111, NAL, Beltsville, MD 20705
Tel: (301) 344-3704

Working Language: English.

Users Policy: Available to public.

Services Available to Public: Library, database searches of AGRICOLA, photocopying, listings of aquaculture-related articles, preparation of bibliographies, interlibrary loans, document delivery.

Description of Collection: Aquaculture documents, rare books, manuscripts, photographs, private library collections.

Subject: Aquaculture.

Publication: List of publications related to aquaculture.

Venezuela

558 Oceanographic Institute of Venezuela (IOV-UDO), Universidad de Oriente, P.O. Box 245, Cumaná

Tel: 093-661817 Telex: 93134 Udons VC

Contact: Director, Biblioteca del Instituto

Working Language: Spanish.

Users Policy: Open to public.

Services Available to Public: Current awareness, abstract service, photocopying, interlibrary loans.

Description of Collection: Information not available.

Subjects: Marine geology, marine sciences, fisheries resources, aquaculture, pollution.

Publications: Accessions list: *Boletín del Instituto Oceanográfico* (every six months); *Lagena* (every six months); *Cuaderno Oceanográfico* (occasional); *Boletín Bibliográfico* (annual) (most subscriptions are sent overseas on an exchange basis).

559 Estación de Investigaciones Marinas de Margarita (EDIMAR), Final Calle Colon, Apartendo 144, Punta de Piedras, Nueva Esparta 6301

Tel: (095) 98236 Telex: 21553

Contact: Librarian

Working Language: Spanish.

Users Policy: Open to public.

Services Available to Public: Literature services, photocopying (fee).

Description of Collection: 9,000 books, 1,200 serial titles.

Subjects: Fisheries biology; physical, chemical and geological oceanography; marine biology; aquaculture; food technology; quality control; microbiology.

Publications: *Memorias de la Sociedad de Ciencias Naturales La Salle*; *Cuadernos Técnicos*; *Documentos de Trabajo Interno*; *Natura.*

Index

Numbers without brackets refer to pages; numbers in square brackets refer to entries in Parts II and III, *not* to page numbers.